DISRUPTIVE ANALYTICS

CHARTING YOUR STRATEGY FOR NEXT-GENERATION BUSINESS ANALYTICS

Thomas W. Dinsmore

Apress®

Disruptive Analytics: Charting Your Strategy for Next-Generation Business Analytics

Thomas W. Dinsmore

ISBN-13 (pbk): 978-1-4842-1312-4   ISBN-13 (electronic): 978-1-4842-1311-7
DOI 10.1007/978-1-4842-1311-7

Library of Congress Control Number: 2016950565

Managing Director: Welmoed Spahr
Acquisitions Editor: Robert Hutchinson
Developmental Editor: Matt Moodie
Technical Reviewer: Robert A. Muenchen
Editorial Board: Steve Anglin, Pramila Balen, Laura Berendson, Aaron Black,
    Louise Corrigan, Jonathan Gennick, Robert Hutchinson, Celestin Suresh John, Nikhil Karkal,
    James Markham, Susan McDermott, Matthew Moodie, Natalie Pao, Gwenan Spearing
Coordinating Editor: Rita Fernando
Copy Editor: Kezia Endsley
Compositor: SPi Global
Indexer: SPi Global
Cover Designer: Isaac Ruiz Soler

Distributed to the book trade worldwide by Springer Science+Business Media New York, 233 Spring Street, 6th Floor, New York, NY 10013. Phone 1-800-SPRINGER, fax (201) 348-4505, e-mail orders-ny@springer-sbm.com, or visit www.springeronline.com. Apress Media, LLC is a California LLC and the sole member (owner) is Springer Science + Business Media Finance Inc (SSBM Finance Inc). SSBM Finance Inc is a Delaware corporation.

For information on translations, please e-mail rights@apress.com, or visit www.apress.com.

Apress and friends of ED books may be purchased in bulk for academic, corporate, or promotional use. eBook versions and licenses are also available for most titles. For more information, reference our Special Bulk Sales–eBook Licensing web page at www.apress.com/bulk-sales.

Any source code or other supplementary materials referenced by the author in this text is available to readers at www.apress.com. For detailed information about how to locate your book's source code, go to www.apress.com/source-code/.

Printed on acid-free paper

Apress Business: The Unbiased Source of Business Information

Apress business books provide essential information and practical advice, each written for practitioners by recognized experts. Busy managers and professionals in all areas of the business world—and at all levels of technical sophistication—look to our books for the actionable ideas and tools they need to solve problems, update and enhance their professional skills, make their work lives easier, and capitalize on opportunity.

Whatever the topic on the business spectrum—entrepreneurship, finance, sales, marketing, management, regulation, information technology, among others—Apress has been praised for providing the objective information and unbiased advice you need to excel in your daily work life. Our authors have no axes to grind; they understand they have one job only—to deliver up-to-date, accurate information simply, concisely, and with deep insight that addresses the real needs of our readers.

It is increasingly hard to find information—whether in the news media, on the Internet, and now all too often in books—that is even-handed and has your best interests at heart. We therefore hope that you enjoy this book, which has been carefully crafted to meet our standards of quality and unbiased coverage.

We are always interested in your feedback or ideas for new titles. Perhaps you'd even like to write a book yourself. Whatever the case, reach out to us at editorial@apress.com and an editor will respond swiftly. Incidentally, at the back of this book, you will find a list of useful related titles. Please visit us at www.apress.com to sign up for newsletters and discounts on future purchases.

The Apress Business Team

To Ann

Contents

Contents

About the Author

Thomas W. Dinsmore is an independent consultant and author who specializes in advanced analytics and machine learning.

In his consulting career, Mr. Dinsmore has served in expert roles for The Boston Consulting Group, PricewaterhouseCoopers, Oliver Wyman, IBM Big Data Solutions, and the SAS Institute. He has also served as Director of Product Management for Revolution Analytics (now a division of Microsoft.)

Mr. Dinsmore has more than 30 years of experience in advanced analytics. He has led or contributed to solutions for AT&T, Banco Santander, Citibank, Dell, J. C. Penney, Monsanto, Morgan Stanley, Office Depot, Sony, Staples, United Health Group, UBS, Vodafone, and many other clients in the United States, Puerto Rico, Canada, Mexico, Venezuela, Brazil, Chile, the United Kingdom, Belgium, Spain, Italy, Turkey, Israel, Malaysia, and Singapore.

Mr. Dinsmore has working experience with most of the leading tools for advanced analytics. He is the co-author of *Modern Analytics Methodologies* (FT Press, 2014) and *Advanced Analytics Methodologies* (FT Press, 2014) and publishes The Big Analytics Blog. He earned an MBA from the Wharton School, The University of Pennsylvania, and a BA from Boston University.

About the Technical Reviewer

 **Robert A. Muenchen** is the author of *R for SAS and SPSS Users* and, with Joseph M. Hilbe, *R for Stata Users*. He is also the creator of r4stats.com, a popular web site devoted to analyzing trends in data science software and helping people learn the R language. Bob is an ASA Accredited Professional Statistician™ with 30 years of experience and is currently the manager of OIT Research Computing Support (formerly the Statistical Consulting Center) at the University of Tennessee. He has taught workshops on research computing topics for more than 500 organizations and has offered training in partnership with DataCamp.com, Revolution Analytics, RStudio, New Horizons Computer Learning Centers, and Xerox Learning Services. Bob has written or co-authored over 70 articles published in scientific journals and conference proceedings, and has provided guidance on more than 1,000 graduate theses and dissertations.

Bob has served on the advisory boards of SAS Institute, SPSS Inc., Intuitics OOD, StatAce OOD, the Statistical Graphics Corporation, and *PC Week Magazine*. His suggested improvements have been incorporated into SAS, SPSS, JMP, STATGRAPHICS, and several R packages. His research interests include statistical computing, data graphics and visualization, text analytics, and data mining.

Acknowledgments

Many thanks to Bob Muenchen of the University of Tennessee; Bob graciously agreed to serve as the technical reviewer of this book and spent many hours reading and commenting on chapter drafts.

Also thanks to Oliver Vagner, Senior Director, Data Analytics at TGI Fridays, and to Professor Dr. Diego Kuonen of the University of Geneva and Statoo Consulting, each of whom provided valuable suggestions and guidance for the book's recommendations to managers.

Thanks also to many other people in the industry who have provided insight into the ideas and topics covered in this book, including Jeremy Achin, Data Scientist and CEO, DataRobot; Bruno Aziza, Chief Marketing Officer, AtScale; Charlie Berger, Oracle; Michael Berthold, President, KNIME.com AG; Arno Candel, Chief Architect, H2O.ai; David Champagne, Principal Software Engineering Manager, Microsoft; Michael Chu, Project Leader, The Boston Consulting Group; Davide Consiglio, Principal, The Boston Consulting Group; Boxuan Cui, Data Scientist, Smarter Travel; David Erdreich, Knowledge Expert, The Boston Consulting Group; Lee Edlefson, Principal Software Engineer, Microsoft; Ali Ghodsi, CEO, Databricks; Mario Inchiosa, Principal Software Engineer, Microsoft; Bill Jacobs, Microsoft; Paul Kent, VP, Big Data, SAS; Josh Klahr, Vice President of Product Management, AtScale; Bill Lehmann, Investor, Bain Capital Ventures; Matthew Madden, Director of Product Marketing, Alteryx; Xiangrui Meng, Software Engineer, Databricks; Ingo Miersewa, Founder and CTO, RapidMiner; Derek Norton, Senior Data Scientist, Microsoft; Thomas Ott, Marketing Data Scientist, RapidMiner; Sean Owen, Director of Data Science, Cloudera; Krishnan Parasuraman, VP Sales, Splice Machine; Zoltan Prekopcsak, VP Big Data, RapidMiner; Peter Prettenhofer, Data Scientist, DataRobot; Dan Putler, Chief Scientist, Alteryx; David Rich, Strategic Advisor and Executive Coach; Razi Razuddin, VP, Strategic Business Development, DataRobot; Vincent Saulys, Senior Director, Advanced Surveillance Development, Financial Industry Regulatory Authority; Joseph Sirosh, Corporate Vice President, Data Group, Microsoft; Dan Steinberg, President, Salford Systems; Ben Strauss, Alteryx and Tableau Evangelist, The Boston Consulting Group; Gregory Todd, Principal, PricewaterhouseCoopers; Deenar Toraskar, Founder, ThinkReactive; Tom Ventura, Global IT Director, The Boston Consulting Group; Reynold Xin, Chief Architect, Databricks; David Wang, Director of Product Marketing, Databricks; and Bill Zanine, Thought Leader, Netezza Analytic Advisory Services, IBM. If I have inadvertently omitted anyone from this list, I apologize in advance.

To my clients, who I cannot name, who have made it possible for me to devote a significant amount of time to this book.

Finally, many thanks to the people at Apress who agreed to publish this book, and to members of the editorial and production staff who contributed to the final product.

Introduction

Disruption: In business, a radical change in an industry or business strategy, especially involving the introduction of a new product or service that creates a new market.

From its birth in 1979, Teradata led the field in data warehousing. The company built a reputation for technical acumen, serving customers like Walmart and Citibank; analysts and implementers alike rated the company's massively parallel databases "best in class." After a 2007 spinoff from NCR, the company grew by double digits.

On August 6, 2012, Teradata released its earnings report for the second quarter. Results excelled; revenue was up 18% and earnings per share (EPS) up 28%. Teradata stock traded at $80, five times its value four years earlier.

"We are increasing our guidance for constant currency revenue growth and EPS for 2012," wrote CEO Mike Koehler.

In retrospect, that moment was Teradata's peak. Over the next three and a half years, the company lost 75% of its market value, as it repeatedly missed revenue and earnings targets. In 2015, Koehler announced a restructuring and sale of company assets; several top executives departed. Finally, after a brutal first quarter earnings report, Koehler himself stepped down in May 2016.

Management blamed many factors for the sluggish sales: long sales cycles, a sluggish economy, and unfavorable currency movement. But worldwide spending on business analytics *increased* during this period and some vendors reported double-digit revenue growth.

Blaming Teradata's struggles on poor leadership would be easy. But the company's growth problems in the last few years are not unique: in the same period, Oracle and IBM suffered declining revenue; Microsoft and SAP failed to grow consistently, disappointing investors; and SAS had to walk back embarrassing projections of double-digit growth, recording low single-digit gains.

In short, while businesses continue to invest in analytics, they aren't buying what the industry leaders are selling.

Meanwhile, a steady stream of innovation creates new value networks in the business analytics marketplace:

Open Source Analytics. With substantial gains in the last several years, open source software makes deep inroads in the analytics community. Surveys show that working data scientists prefer open source R and Python over commercial software. Technology leaders like Oracle, IBM, and Microsoft rush to get on the open source bandwagon.

Hadoop and its Ecosystem. As Hadoop matures, it competes successfully with data warehouse appliances, even displacing them. Technology consultant Gartner estimates that 42% of all enterprises now use Hadoop. A few years ago, data warehousing vendors laughed at Hadoop; they aren't laughing today.

In-Memory Analytics. As the cost of memory declines, fast and scalable high-performance analytics are within reach for any organization. Adoption of open source Apache Spark, an open source project for scalable in-memory computing, increases exponentially. With more than a thousand contributors, Spark is the most active open source project in Big Data.

Streaming Analytics. Organizations face a growing volume of data in motion, driven in part by the Internet of Things (IoT). Today, there are no less than six open source projects for streaming analytics in the Apache ecosystem. In-memory databases position themselves as streaming engines for hybrid transactional/analytical processing (HTAP).

Analytics in the Cloud. When Amazon Web Services introduced its Redshift columnar database in 2012, it lacked many of the features available in competing data warehouses. For many businesses, however, Amazon offered a compelling value proposition: "good enough" functionality, at a fraction of the cost of a Teradata warehouse. The leading cloud services all report double-digit revenue growth; Gartner estimates that 44% of all businesses use the cloud.

Deep Learning. Cheap high-performance computing power makes Deep Learning practical. NVIDIA releases its DGX-1 chip for Deep Learning, with the power of 250 servers; Cray announces its Urika-GX appliance with up to 1,728 cores and 35 terabytes of solid-state memory. Meanwhile, Google releases its TensorFlow framework to open source and declares that it uses Deep Learning in "hundreds" of applications.

Self-Service Analytics. With an easy-to-learn user interface and robust connectors to data sources, Tableau turns the business intelligence software industry upside down and grows its revenues tenfold while established Business Intelligence vendors struggle to adapt. Other startups position themselves to bring the self-service model to other disciplines, such as OLAP and machine learning.

This is not another book that hypes Big Data. Petabytes of data are worthless unless they answer a business question; the tsunami of data produced by the digital economy is simply a fact of life that managers must address. Whether you manage a multinational or drive a truck, your business produces more data than ever; you will either use it or discard it, but one way or the other, you must make an informed decision.

In a disrupted business analytics market, managers must focus ruthlessly on needs for insight, then build systems and processes that satisfy those needs. Understanding the innovations described in these chapters is a step toward that end, but the focus must remain on the demand for insight and the value chain that delivers it.

Innovations do not spring fully formed from the mind of an inventor; they are the end result of a long process of tinkering. Many of the most significant innovations we describe in this book are more than 50 years old; they emerge today for various reasons, such as the long-run decline of computing costs. We present a historical perspective at several points in this book so the reader can distinguish between that which is really new and that which is simply repackaged and rebranded.

In the middle chapters of this book, we present a survey of a key innovation in business analytics. These chapters include detailed information about available software products and open source projects. In general, we do not cover offerings from industry leaders, under the premise that these companies have ample marketing budgets to build awareness of their products.

We close the book with a handbook for managers: specific strategies to profit from disruptive innovation. Some of these strategies may seem radical; if this disturbs you, put this book down—it's not for you. But if you are ready to embrace disruptive innovation, and profit by it, read on.

Fundamentals

Disruption in the Analytics Value Chain

The analytics business is booming. Technology consultant IDC estimates[1] total spending for analytic services, software, and hardware exceeded $120 billion in 2015; through 2019, IDC forecasts that spending will increase to $187 billion, an 11% compound annual growth rate[2].

So, if analytics is such a hot field, why are the industry leaders struggling?

- Oracle's cloud revenue growth[3] fails to offset declining software and hardware sales[4].

- SAP's cloud revenue grows, but total software revenue is flat[5].

- IBM reports[6] 16 straight quarters of declining revenue. Mass layoffs ensue[7].

---

[1] https://www.idc.com/getdoc.jsp?containerId=IDC_P33195
[2] http://www.cio.com/article/3074238/analytics/big-data-and-analytics-spending-to-hit-187-billion.html
[3] http://www.forbes.com/sites/laurengensler/2016/03/15/oracle-third-quarter-earnings/#286720039d5d
[4] http://investor.oracle.com/financial-news/financial-news-details/2016/Oracle-Reports-GAAP-EPS-of-050-Non-GAAP-EPS-of-064-Without-the-Effect-of-US-Dollar-Strengthening-Both-Would-Have-Been-4-Cents-Higher/default.aspx
[5] http://go.sap.com/docs/download/investors/2016/sap-2016-q1-statement.pdf
[6] https://www-03.ibm.com/press/us/en/pressrelease/49554.wss
[7] http://fortune.com/2016/05/20/ibm-layoff-employees-may/

© Thomas W. Dinsmore 2016
T. W. Dinsmore, *Disruptive Analytics*, DOI 10.1007/978-1-4842-1311-7_1

- Microsoft underperforms[8] analysts' expectations despite 120% growth in Azure cloud revenue.

- Predictive analytics leader SAS reports[9] five years of low single-digit revenue growth; EVP departs[10].

- Data warehousing leader Teradata shuffles its leadership team after four years of declining product revenue[11].

Product quality is not the problem. Each company offers products that industry analysts rate highly:

- Forrester and Gartner both[12] recognize[13] IBM, SAS, SAP, and Oracle as leaders in data quality tools.

- Gartner rates[14] Oracle, SAP, IBM, Microsoft, and Teradata as leaders in data warehousing.

- Forrester rates[15] Microsoft, SAP, SAS, and Oracle as leaders in agile business intelligence.

- Gartner recognizes SAS and IBM as leaders in Advanced Analytics[16].

The answer, in a word, is *disruption*[17]. Powerful forces are rearranging the industry:

- Digital transformation of the economy and rapidly declining storage costs produce a data tsunami.

- The number of data sources is exploding. Data sources are everywhere: on-premises, in the cloud, in consumers' pockets, in vehicles, in RFID chips, and so forth.

---

[8]http://www.reuters.com/article/us-microsoft-results-idUSKCN0XI2NG
[9]http://www.sas.com/en_us/company-information.html#stats
[10]http://www.newsobserver.com/news/business/article48668040.html
[11]http://www.mydaytondailynews.com/news/news/teradata-leadership-change-comes-as-company-strugg/nrHwg/
[12]http://www.sas.com/en_us/news/analyst-viewpoints/forrester-names-sas-leader-in-data-quality-solutions.html
[13]http://www.sas.com/en_us/news/analyst-viewpoints/gartner-names-sas-leader-in-data-quality-tools.html
[14]http://www.gartner.com/doc/reprints?id=1-2ZFVZ5B&ct=160225&st=sb
[15]http://www.forrester.com/pimages/rws/reprints/document/116447/oid/1-SFDMEH
[16]http://www.sas.com/en_us/news/analyst-viewpoints/2016-gartner-magic-quadrant-advanced-analytics.html
[17]http://blogs.forrester.com/brian_hopkins/15-11-03-ibm_and_teradata_a_tale_of_two_vendors_struggle_with_disruption

- Data *governance* is complicated by decentralized data *ownership* as functional executives control an increasing share of technology spending.

- The open source software business model offers an increasingly attractive alternative to commercial software licensing.

- Increasingly, the Hadoop ecosystem displaces conventional data warehousing; R and Python displace commercial analytic software.

- The elastic business model made possible by cloud computing undercuts conventional software licensing and provisioning.

- Widely available and inexpensive computing power make computationally intensive techniques like Deep Learning practical.

Consider what has happened to Teradata. Late in 2012, the company started missing sales targets; in early 2013, it stunned investors by reporting an absolute decline in sales. Management offered excuses; Wall Street punished the stock, driving it down by half in the face of an overall bull market.

From 2013 through early 2016, Teradata continued to miss sales and earnings targets; Wall Street drove the stock price down to a fraction of its 2012 peak. While it is tempting to blame the problem on poor leadership, Teradata's persistent failure to forecast its own sales and earnings indicates something amiss. The world changed; the value networks created in Teradata's rise to leadership no longer exist; the mental models managers used to understand the market no longer work.

Disruptive Innovation

Clayton Christensen of the Harvard Business School outlined[18] the theory of disruptive innovation in 1997. We summarize the theory briefly; for an extended discussion, read Christensen's book:

- Industries consist of value networks, collections of suppliers, channels, and buyers linked by relationships.

- Innovations disrupt industries when they create a new value network.

---

[18]Christensen, Clayton M. (1997), The innovator's dilemma: when new technologies cause great firms to fail, Boston, Massachusetts, USA: Harvard Business School Press, ISBN 978-0-87584-585-2.

- Not all innovations are disruptive. Many innovations are introduced by market leaders to sustain a competitive position.

- Disruptive innovations tend to be introduced by outsiders.

- Purely technological innovation is not disruptive; what matters is the *business model* enabled by the new technology.

Christensen identified two forms of disruption. *Low-end disruption* occurs when industry leaders enhance products faster than customers can assimilate the enhancements; the disruptor enters the market with a "good enough" product and a better value proposition. The disruptor's innovation makes it possible to serve customers at a lower cost than the industry leaders can deliver.

New market disruption takes place when the disruptor innovates in ways enabling it to serve customers that are not served by the industry leaders.

In this book, we discuss two kinds of disruption. The first is disruptive innovation *within* the analytics value chain (a concept we explore later in this chapter). The second is industry disruption *by* innovations in analytics.

There are many examples of disruption *within* the analytics value chain:

- Hadoop disrupts the data warehousing industry from below. Hadoop does not do everything a relational database can do; but it does just enough to offer an attractive value proposition for the right use cases. When first introduced, Hadoop's capabilities were quite limited relative to data warehouse appliances. But Hadoop's flexibility and low cost were highly attractive for applications that did not need the performance and features of a data warehouse appliance. While established vendors struggle to maintain flat and declining revenue, Hadoop distributors grow at double-digit rates.

- Tableau virtually created the market for agile self-service discovery. Tableau has no charting and visualization features not already available in mainstream business intelligence tools. But while business intelligence vendors targeted the IT organization in large enterprises and continuously added features, Tableau targeted the end user with a simple, easy to use, and versatile tool. As a result, Tableau has increased its revenue tenfold in five years, leapfrogging over many other BI vendors.

Examples of disruption *by* analytics are less prevalent, but they do exist:

- General-purpose credit scoring introduced by Fair, Isaac and Co. in 1987 virtually created a national market in credit cards. Previously, banks issued credit cards to their local customers, with whom they had an established relationship. Uniform credit scoring enabled a few large issuers to identify creditworthy customers in the general population, without a prior relationship.

- When the U.S. Securities and Exchange Commission authorized electronic trading in regulated securities in 1998, market participants quickly moved to develop algorithms that could arbitrage between markets, arbitrage between indexes and the underlying stocks, and exploit other short-term opportunities. Traders that most effectively deployed machine learning for electronic trading grew at the expense of other traders.

The relative importance of the two kinds of disruption depends on the reader's perspective. Disruption within the analytics value chain is pertinent for readers who plan to invest in analytics technology for their organization. Technologies at risk of disruption are risky investments; they may have abbreviated useful lives, and their suppliers may suffer from business disruption. Taking a "wait-and-see" attitude toward disrupted technologies makes good sense, if only because prices will likely decline in the future.

For startups and analytics practitioners, disruption *by* analytics is key. To succeed, startups must disrupt their industries. Using analytics to differentiate a product is a way to create a disruptive business model or to create new markets.

To understand disruptive *analytics*, we must first understand the current state of analytics and its drivers. In the remainder of this chapter, we present a discussion of what drives the demand for analytics, and an overview of the analytics value chain. We close the chapter with an outline of the rest of the book

The Demand for Data-Driven Insight

The key to survival in a disrupted world is to ruthlessly re-examine business processes, working backward from a problem.

Analytics is *the systematic production of useful insight from data*. In business, people use insight to solve one of five core problems:

- Develop a business strategy.

- Manage a business unit.

- Optimize a business process.

- Develop products and services.

- Differentiate products and services.

Each of these problems needs a different kind of insight, whose delivery requires distinctive people, processes, and tools.

Developing a Business Strategy

We define "strategy" narrowly to mean choices made by the top leadership of an organization: the "C-Suite". Many people may participate in the development of strategy, but in every organization, the buck stops somewhere. Strategic analytics are any analytics that support strategic decisions.

What makes an issue "strategic?" Strategic questions and issues have four distinct characteristics:

- The stakes are high; there are major consequences that depend on making the right choice. (Otherwise, the issue will be delegated.)

- The issue falls outside of existing policy; no established rule enabling decisions at a lower level. (There may be a conflict of policies, or the situation may be unprecedented.)

- Strategic issues are non-repeatable; in most cases, the organization addresses a strategic question once and never again. (Repeatable decisions are handled at lower levels through policy.)

- There is no clear consensus about the best choice. (If everyone agrees on the best choice from the outset, there is no need for analysis).

Examples of strategic topics include:

- Technology or product investments

- Mergers and acquisitions

- Business portfolio restructuring

- Business reorganization

- Branding, rebranding, and product positioning

- Crisis management

Since the stakes are high for strategic analytics, so is the sense of urgency; some decisions, like merger proposals, may be strictly bounded in time. Crises provoked by product failure, natural disasters, or other issues may have actual life and death implications.

Deliverables for strategic analysis include reports, charts, visuals, and presentations. Owing to the high stakes of the decision, executives closely scrutinize the presented analysis. Analysis must be "bullet-proof," especially if the results do not square with leadership's prior beliefs. The methods used to produce the analysis must be clear.

Due to the ad hoc and non-repeatable nature of strategic analytics, enterprise data warehouses (EDWs) play at most a supporting role. In most cases, the data in EDWs is internal and supports existing processes with well-defined requirements. The data needed to support strategic decisions often comes from external sources, may be difficult to access, and may be needed only once.

Enterprises frequently engage outside consultants to deliver strategic analysis. While organization insiders may have no experience in a particular type of problem, outside experts have deep experience with similar problems. Firms also prize consultants' independence and neutrality, since strategic decisions require resolving competing internal interests.

Managing a Business Unit

Managerial analytics support decisions a level down in the organization from top leadership. At this level, needs for analysis link to specific functions, such as Treasury, Product Management, Marketing, Merchandising, Operations, and so forth.

There are three distinct applications for managerial analytics:

- Performance measurement
- Performance optimization
- Business planning

Performance measurement is the sweet spot for enterprise business intelligence (BI) systems. BI is highly effective when the data is timely and credible, reports are easy to use, and metrics align with business objectives. Most organizations want to measure business units in a consistent manner, so they ordinarily implement reporting systems centrally rather than letting business unit managers measure themselves.

Metrics tell the manager which entities (e.g., brands, products, campaigns, stores, and sales reps) performed well and which entities performed poorly. Optimization delivers guidance on how to improve or optimize performance by shifting budget investments. Marketing mix analysis, for example, estimates the revenue impact of spending on different channels and programs, so the organization can shift the marketing budget to its most productive uses.

Finally, business planning is a process of goal setting and goal alignment across functions, where the manager justifies operating and capital spending. In large organizations, the business planning process is highly templated and structured. Forecasting is an important tool for business planning.

Deliverables for managerial analysis are similar to strategic analysis. Detailed analysis and forecasts may be in the form of queryable "cubes" or interactive tools.

Optimizing Business Processes

Optimization at this level is much more granular than optimization for functional leadership. In marketing, for example, the CMO needs summary information about the effectiveness of all major programs; the CMO's optimization problem requires shifting budget among programs. The program manager, on the other hand, seeks to optimally match programs, value propositions, and creative treatments to individual customers and customer segments.

There are many ways that analytics can optimize a business process. Examples include:

- Automated decision engines
- Targeting and routing systems
- Operational forecasting systems

Automated decision engines apply consistent rules designed to balance risks and rewards. Embedded analytics help optimize criteria and ensure that decision rules reflect actual experience. Decision engines are faster than human decision-makers and make better decisions. Examples include payment authorization systems and credit approval systems.

Targeting and routing systems evaluate the characteristics of an incoming message or request and direct it to the appropriate agent or subsystem. Analytics extract essential information from the request, eliminating manual evaluation and triage. Examples include e-mail routing systems in customer service operations and SAR investigation routing systems in bank anti-money-laundering systems.

Operational forecasting systems project key metrics that affect operations, enabling the organization to align resources accordingly. Analytics leverage historical data to detect traffic patterns and shift resources to locations or shifts where they are most needed. Examples include retail staffing systems that plan shifts based on expected floor traffic, and police patrol routing systems that direct officers to projected high-crime areas.

Analytics that optimize business processes are ordinarily embedded in production systems, and usually must operate in real time. This implies a need for streaming analytics, which we cover in Chapter Six. Analytic deliverables are machine-consumable models implemented in software.

Developing Products and Services

The development process in organizations runs the gamut from creative brainstorming to formal scientific research, as in pharmaceutical laboratories, to "skunk works" prototyping. As such, the range of possible analyses is extremely broad. Developmental analytics fall into two broad categories:

- Analytics for generating hypotheses
- Analytics for testing hypotheses

Managers perform or commission hypothesis-generating analysis to identify unmet consumer needs or gaps in existing products. This can include activities like analyzing external data consumer surveys and consumption data; analyzing operational data; or evaluating clinical reports of treatment for a certain disease.

At a later stage in the product development process, managers test hypotheses about specific product concepts, prototypes, or small production run. Analysis at this stage can include analyzing clinical trial data to determine the efficacy of a drug; and analyzing test market data to assess the value of a product feature, or similar activities.

For practitioners, specialized domain expertise dominates purely analytical skills in this area. (One would not expect a biomedical specialist who specializes in Parkinson's disease to easily switch to developing trading algorithms for a hedge fund.) Analytic processes must be highly flexible and agile, adapting to the particular problem at hand based on the product development cycle.

Differentiating Products and Services

We distinguish between analytics that *support* product development, and analytics that *are* the product, or embedded analytics.

For the previous four use cases, the "consumer" of insight is inside the organization—a top executive, functional manager, process participant, or product developer. Increasingly, however, analytics provide insight to end consumers outside of the organization. In these cases, analytics differentiate the product and make it stand out in the marketplace.

As the volume and variety of information available to consumers explodes, insight itself becomes a valued commodity. In this world, the most powerful analytic applications aren't often viewed as analytics. Is Google an analytic application? Google uses analytics technology, including content analytics and graph analytics, and it produces a particular kind of insight.

Online retailing's ability to carry a vastly larger number of unique items than brick-and-mortar retailers creates a shopping problem for consumers; with so many items from which to choose, what should we buy? Recommendation engines, which use machine learning to optimal products for an individual customer, are widely used. Most readers will be familiar with some salient examples, all of which use machine learning:

- Facebook leverages a user's profile and likes to optimize the news feed.

- Streaming video sites like Netflix leverage the user's ratings and other information to personalize recommendations.

- Tinder pairs users based on profile and "swipes."

- Amazon.com uses data-driven similarity ratings to display products that are compatible to what a user has selected.

- Spotify leverages a user's prior preferences and content analytics to optimize the music stream.

Success in embedded analytics is a matter of software engineering; the end product must be tightly packaged for reliability and usability; in most cases it must operate in real time.

The Analytics Value Chain

Once we understand the demand for insight, we can define a value chain. The analytics value chain begins with data and ends with insight, progressively transforming data from low value to high value in a sequence of steps, as Figure 1-1 shows.

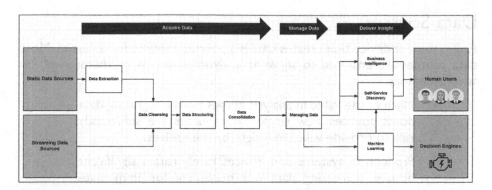

Figure 1-1. The analytics value chain

Of course, it's possible to define the value chain at a much finer level of detail than we show here. At a high level, the analytics value chain includes three major components: steps that acquire data, steps that manage data, and steps that deliver insight. Delivering insight to human or machine users is the critical link in the chain; a system that successfully acquires and manages data but does not deliver insight has failed.

Acquiring Data

All data comes from an original source; capturing data from sources is the first step in the value chain. The processes that capture and manage data are variously called Extract/Transform/Load (ETL), Data Integration, or Master Data Management (MDM). ETL refers to the physical movement of data; data integration addresses the challenge of consolidation across sources; and MDM addresses governance and administration of the process. Commercial vendors offer software to manage data flows through the value chain, cleanse the data, and load it into a analytic datastore. According to IDC, Informatica leads the commercial market, followed by SAS and IBM. Talend, Pentaho, and JasperSoft offer open core software, and Apache NiFi is a full open source project to manage data flows. (We discuss open source software business models in Chapter Three.)

Data Sources

Any system or device that creates data is a potential source for analytics. Most data sources are unsuited to serve as analytic platforms by themselves, for several reasons:

- Much of the value in analysis comes from integrating data *across* sources. Few single data sources are sufficiently rich to provide valuable insight by themselves.

- Production systems and devices rarely retain significant history, truncating data not necessary for immediate transaction processing needs.

- Production systems and devices are usually designed to support transaction workloads and not analysis workloads.

Data sources are either static or streaming. Static data sources accumulate new data until a user requests data through a query or extract operation. Streaming data sources continuously publish data, "pushing" data to subscribers.

Data Extraction

The first step in the value chain is to "extract" data from one or more static source systems. While conventionally called "extraction," in most cases data is *copied* rather than extracted.

For streaming data sources, this step is not necessary.

The organization that manages the production system (e.g., the IT organization) rarely permits free access to production systems, for two reasons:

- The extract operation cannot interfere with transaction processing.

- Production systems often contain sensitive information that must remain under data security protocols.

Hence, the organization that owns the production system generally controls the extract process and implements the procedure under a service level agreement.

At the beginning of the data warehouse era in the 1980s, the IT organization "owned" virtually all of the prospective data sources for analysis. As we discuss in Chapter Two, the digital transformation of business processes leads to an increasing share of technology spending controlled by functional executives. This, in turn, means that functional executives control the data sources as well.

Another radical change from the early days of data warehousing is the increased use of cloud computing and Software-as-a-Service platforms for production systems. This means that data sources are less and less likely to be physically located on-premises.

Data Cleansing

Data from source systems may be "dirty": it may be inaccurate, incomplete, or erroneous. Data cleansing software scans incoming data and checks to see if items satisfy validity tests and are internally consistent. When the software finds an exception, it either force cleans the item or queues it to an exception file for human analysis.

Data cleansing ensures that data conforms to business logic, but it does not ensure accuracy. Verifying accuracy requires comparison to a reference value, which can only exist under lab conditions.

Few organizations have the resources to consistently research data cleaning exceptions. In practice, most issues in data are discovered by actual users with subject matter expertise.

Cleaning data in the analytics value chain violates the third of quality guru W. Edwards Deming's 14 principles[19] of business transformation:

> *Cease dependence on inspection to achieve quality. Eliminate the need for massive inspection by building quality into the product in the first place.*

Rather than inspecting cars at the end of an assembly line and scrapping the ones that fail, it makes much better sense to design quality into the process and build high-quality cars. Similarly, it is much smarter to build data quality directly into the source systems that generate data than it is to trap and correct errors farther down the chain.

Data Structuring

We avoid use of the term "unstructured data" in this book. All data has structure. Some data has structure that is not yet known, and some data is difficult or impossible to map into the entity-relational (ER) framework that is the foundation of relational databases. Examples of such data include text, audio, video, images, and log files.

In conventional data warehousing practice, the data consolidation process is also a standardization process. This resolves differences in data structure, so that all data conforms to a unified data model—otherwise, it can't be consolidated.

---

[19]Deming, W. Edwards (1986). *Out of the Crisis*. MIT Press.

Some data is structured from inception, because the source system that produces it uses a relational database for storage. If the data model of the source system aligns with the data model that governs structured data in the analytics value chain, the data can be used directly.

However, even if the source data is structured, it may conform to a different data model than the analytics value chain. In this event, the data must be restructured or mapped into the desired data model.

Some data is semi-structured: the data itself includes information about its structure. In this case, the organization must decide whether to structure the data prior to storing it, or to simply catalogue it and defer structuring until it is used.

Log files can be parsed and structured with special tools. Text, audio, video, and images generally cannot be structured into an ER framework; however, machine learning tools (discussed later in the chapter) can scan content to identify duplicates or classify it into categories.

Data Consolidation

For insight, organizations consolidate information from many data sources. In most cases, data sources lack a common data structure, for several reasons:

- Data sources may include systems and devices from different manufacturers that produce data to different standards.

- Large organizations may have many systems implemented at different times or acquired in mergers and acquisitions.

- Source systems and devices may produce data that is difficult to map into a relational data model, such as log files.

More recently, with the growth of text, images, audio, and video data, standardization is difficult or impossible. In this environment, "consolidation" simply means the aggregation of files, with structuring postponed to the query phase of analysis.

Managing Data

An analytic datastore is *any repository that holds data collected from original sources in a format that facilitates analysis.* Every analytics value chain has one or more intermediate datastores variously called data warehouses, data marts, and data lakes. In large, mature organizations, there may be many analytic datastores.

Every analytic datastore should serve three primary purposes:

- Accumulating history
- Collecting data across sources
- Cataloguing and organizing the data

Accumulating history is a key function of the analytic datastore, since primary data sources generally do not perform this function. Data not retained in an analytic datastore is simply lost.

As noted previously, the production of insight generally requires combining data from multiple sources. Consolidating data in a single repository adds value by saving time for the analyst, just as a supermarket saves time for the food shopper, who otherwise would have to make separate trips to the butcher, produce store, bakery, and so forth.

Data that is not catalogued is lost. Imagine an enormous library without an index, where books are simply stacked on shelves at random. Who would use such a library? In an analytic datastore, data is indexed and searchable, and its lineage is documented.

Analytic datastores must also support the organization's data security policies. Since data preservation is a key priority, they must have backup, restore, and disaster recovery capabilities.

Theorists engage in extended debates about the definition of terms like *data warehouse*, *data mart*, and *data lake*; they also debate the relative merits of each architecture. The debates are academic and a waste of time; no organization can choose an optimal architecture for an analytic datastore in the abstract, without reference to an actual end user.

Of course, when data is created we don't necessarily know how end users will want to produce insight. In the absence of firm requirements, organizations should simply catalogue and archive data in atomic form at the lowest possible cost, deferring more complex data integration until clear business cases emerge.

Oracle leads the commercial market for software to build analytic datastores, followed by IBM, Microsoft, Teradata, and SAP. The top five vendors control 80% of the market, according to IDC. In Chapter Four, we discuss the Hadoop ecosystem, an open source alternative to the leading commercial platforms.

Delivering Insight

Acquiring and managing data is an essential part of the analytics value chain, but delivering insight produces the most value. Figure 1-2 shows worldwide business analytics software spending forecast by IDC by high-level categories; about two-thirds of all projected spending is for software that delivers insight to end users. This includes spending on query, reporting, and analysis tools; advanced and predictive analytics tools; spatial analytics tools; content analytics tools; and performance management and analytic applications.

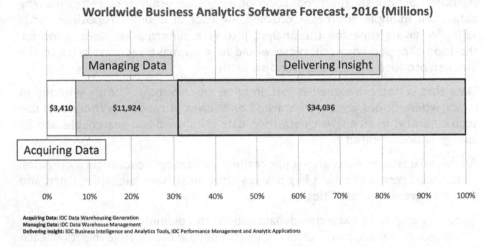

Worldwide Business Analytics Software Forecast, 2016 (Millions)

Acquiring Data: IDC Data Warehousing Generation
Managing Data: IDC Data Warehouse Management
Delivering Insight: IDC Business Intelligence and Analytics Tools, IDC Performance Management and Analytic Applications

Figure 1-2. *Software spending in the analytics value chain*

In the sections that follow, we survey three major categories of tools and processes that produce insight: business intelligence, self-service discovery, and machine learning.

Business Intelligence

We can resolve many business issues with simple quantification:

- How many cases of Product X did we sell in Region Y?

- How much did each of our sales representatives sell in the first quarter?

- What was our sales volume by category in each of the past four quarters?

In each case, the question can be addressed by aggregating facts into measures by dimensions. For example, in the first example, the facts are sales transactions; the measure is "number of cases"; and the dimensions are Product and Region.

For questions in this form, queries against relational databases with Structured Query Language (SQL) deliver the needed answer. In Chapter Two, we discuss SQL in its historical context.

Most business users prefer to interact with data through business intelligence (BI) tools rather than directly through SQL. Business intelligence tools offer a graphical user interface and "business-friendly" views of the data. Behind the scenes, however, BI tools generate SQL or MDX, a competing standard for queries.

Reports are formatted views of data, typically containing many individual items. Typically in the form of tables or cross-tabulations reports contain primary measures with calculated statistics. For example, a report showing the number of sales transactions and their dollar value by region can also show statistics calculated from those measures, such as the average value of a transaction by region, or the percentage distribution of sales across regions.

Dashboards are collections of individual measures, reports, and graphical displays that summarize key metrics. Organizations build predefined dashboards to support ongoing initiatives; for example, a customer service operation might develop a dashboard that summarizes many key service quality metrics.

There are three principal applications for quantification in an enterprise. The first of these is performance measurement. After taking an action, managers want to measure its success—or lack thereof. Moreover, managers have an ongoing interest in the performance of their domain, under the premise that "you can't manage what you don't measure"[20].

Managers place a premium on accurate, consistent, and timely performance reporting based on well-defined metrics. They also value metrics with a clear tie to the organization's goals and objectives. Business intelligence tools perform very well for performance measurement, as they excel at delivering consistent and repeatable metrics to a large audience.

The second application is interactive discovery to support program and product development. For this application, questions are less well defined than for performance measurement; the answer to one question raises many other questions, analogous to peeling an onion.

Conventional business intelligence tools perform less well for this application than they do for performance measurement; they tend to be relatively inflexible, better suited to production reporting than agile discovery. OLAP tools designed for dimensional analysis are a little more flexible than reporting tools, but business users with high needs for interactivity may work directly with SQL.

---

[20] http://management.about.com/od/metrics/a/Measure2Manage.htm

The third application—business planning—requires forecasting as well as historical analysis. Most business intelligence tools support simple time series analysis, which is sufficient for many managers. In other cases, managers may integrate historical data from a business intelligence tool with forecasts developed by specialists, combining the two sets of values in a spreadsheet or presentation tool.

While queries, reports, and dashboards are powerful tools, they are limited to low-dimensional problems where the question can be addressed within the framework of facts, measures, and dimensions. Dimensionality is a key issue for these tools. An analyst can easily work with reports showing data in one, two, or three dimensions; with graphics, four and even five dimensions are feasible. With more than five dimensions to consider, the analyst must break the problem into separate low-dimensional analyses; the number of possible combinations rises exponentially as the number of dimensions increases.

Self-Service Discovery

Conventional business intelligence tools are too inflexible to support interactive discovery. Self-service discovery tools, on the other hand, are ideally suited to this application.

While sometimes called "visualization" tools, the charting and graphics capabilities of tools in this category are no better than many other analytic software packages on the market. These tools have three outstanding features:

- Simplified user interface that is easy to learn and use

- Basic charting and graphics functionality that aligns well with what most managers need

- Flexible "back end" that simplifies connection to many different data sources

Among commercial vendors, Tableau Software and Qlik are the market leaders. Microsoft PowerBI and SAP Lumira also score very well in analyst evaluations[21].

We cover self-service analytics in more detail in Chapter Nine.

Machine Learning

Machine learning is a set of algorithms and a discipline that governs how to use them. Machine learning identifies patterns in data that are inaccessible to a human user and produces output in human or machine-consumable form.

---

[21]http://www.forrester.com/pimages/rws/reprints/document/116447/oid/1-SFDMEH

There are many techniques for machine learning, hundreds of algorithms and thousands of software implementations of those algorithms. We discuss machine learning at a managerial level here and in in Chapter Eight. For technical treatment of the subject, there are many excellent books[22] on the machine learning discipline as a whole, and on individual techniques.

Data scientists distinguish between techniques for *supervised* and *unsupervised* learning. Supervised learning techniques require training data where the outcome we wish to predict is known. For example, if we want to predict which prospects will respond to a campaign, we need data for prospects targeted by the campaign showing whether or not they responded.

Supervised techniques provide powerful tools for *prediction* and *classification* problems. In classification problems, the outcome we wish to predict is categorical, such as response or no response. In prediction problems, the outcome we wish to predict is an amount, such as a customer's future spending.

Frequently, however, we do not know the "ultimate" outcome of an event. For example, in some cases of fraud, we may not know that a transaction is fraudulent until long after the event. In this case, rather than attempting to predict which transactions are frauds, we might want to use machine learning to identify transactions that are unusual and flag these for further investigation. We use unsupervised learning when we do not have prior knowledge about a specific outcome, but still want to extract useful insights from the data[23].

While some machine learning techniques tend to consistently outperform others, it is rarely possible to say in advance which one will work best for a particular problem. Hence, most data scientists prefer to try many techniques and choose the best model. For this reason, high performance is essential, because it enables the data scientist to try more options and build the best possible model[24].

The potential applications for machine learning in organizations are highly diverse. For supervised learning, the three most common use cases are:

Prediction. Estimating the incidence or value of a measure that is unknown because it takes place in the future. For example, a bank seeks to predict the odds that a borrower will repay a loan during its term when evaluating an application; a retailer seeks to predict store traffic next week when scheduling staff. The temporal dimension, the element of time, plays a key role.

---

[22]For example, Hastie, Tibshirani and Friedman, *The Elements of Statistical Learning*, Springer (2011); Provost and Fawcett, *Data Science for Business*, O'Reilly Media (2013).
[23]http://www.infoworld.com/article/3010401/big-data/machine-learning-a-practical-introduction.html.
[24]http://university.h2o.ai/business-101/downloads/practical-guide-to-machine-learning.pdf.

Organizations use prediction to support operational decisions on a large scale. Modern credit card operations, for example, are only possible because issuers can make rapid decisions to approve credit lines and authorize transactions. Such operations depend on predictive models developed with machine learning.

Inference. Estimating the odds or amount of an unknown measure that is not a future event. For example, a retailer seeks to determine the ethnicity of its customers through analysis of surnames, street addresses, and purchase behavior.

Attribution. Disaggregating the contribution of many factors to a desired outcome. For example, an ecommerce vendor seeks to determine how ad exposures impact sales; a sports team seeks to measure the contribution of each player to winning games. Executives rely on attribution for managerial and strategic decisions to allocate budgets, continue or discontinue programs, and similar decisions.

There are numerous applications for machine learning in content analytics:

- Text processing applications extract features from text for visualization or inclusion in predictive models.

- Machine learning can match documents to detect duplicates or identify plagiarism.

- Image processing can classify images into categories, detect malignant tumors in cancer screenings, and so forth.

While SAS and IBM combined control[25] a little less than 50% of the commercial software market in machine learning, the market as a whole is less concentrated than elsewhere in business analytics. This is largely due to rapid innovation in machine learning, and overall rapid expansion in the number of potential applications.

Cloud-based services from Amazon Web Services, Microsoft, and Google have the potential to disrupt the established leaders; we discuss them in Chapter Seven. Open source offerings like R, Python, Spark, and H2O are gaining users at the expense of commercial vendors; we discuss them in Chapters Three, Five, and Eight.

---

[25]http://www.sas.com/content/dam/SAS/en_us/doc/analystreport/idc-apa-software-market-shares-108013.pdf

Overview of the Book

Chapter Two is a short history of business analytics. It covers the last 50 years of innovation in analytics, to provide context for innovations currently impacting the analytics value chain.

In Chapter Three we cover the open source business model, including licensing and distribution. Today, there are open source options everywhere in the value chain, and enterprise adoption is on the rise.

Chapter Four covers the Hadoop ecosystem. Due to the importance of SQL processing in analytics, we also cover open source SQL engines in this chapter.

In Chapter Five, we document the rapidly declining cost of computer memory and the corresponding rise of large-scale in-memory computing, including in-memory databases and Apache Spark.

Chapter Six is a survey of streaming analytics. We include a brief history of streaming analytics for context, and introduce the reader to open source streaming platforms.

Cloud computing and the elastic business model is disrupting the software industry. The elastic business model is especially appropriate for analytics. We survey analytics in the cloud in Chapter Seven.

We briefly summarized machine learning in this chapter. In Chapter Eight, we discuss recent innovations in machine learning, with special emphasis on Deep Learning.

In Chapter Nine, we cover self-service analytics. Tableau's self-service model is one of the best examples of disruption in the analytics value chain.

Finally, in Chapter Ten, we offer a manager's handbook for disruptive analytics. We survey the key requirements—people, process, and tools—needed to build a platform for disruption.

A Short History of Analytics

Developing the Analytics Value Chain

This chapter is a short history of analytics in the era of modern enterprise computing. The story of business analytics over this period is one of progressive development from fully integrated but closed systems, to open, modular, and increasingly complex processes.

We divide this chapter into three broad sections:

- Before the data warehouse, when unified and closed systems supported a complete analytics value chain for a narrow audience.

- The era of the data warehouse, when open standards enabled the decoupling of ETL, data management, and business intelligence.

- Key trends in the economy disrupting established value chains today.

Within each period, we discuss the separate development of data warehousing, business intelligence, and predictive analytics. We also cover, in context, two excellent examples of disruptive analytics in action.

© Thomas W. Dinsmore 2016
T. W. Dinsmore, *Disruptive Analytics*, DOI 10.1007/978-1-4842-1311-7_2

Before the Data Warehouse

Four themes characterize analytics in the years prior to the introduction of the first data warehouse in 1984:

- Computing was expensive by today's standards, and relatively few analytic use cases met the threshold for investment.

- With the introduction of the IBM System/360, enterprise data expanded rapidly, in a jumble of complex and proprietary formats.

- Business intelligence was rudimentary, expensive, and siloed.

- Statistics and machine learning were largely academic tools used at scale by a few firms with close ties to research, such as the pharmaceutical and insurance industries.

In 1969, the IBM System/360 Model 75 mainframe computer cost $3.5 million ($23 million in 2016 dollars). That computer had megabytes of memory and could perform several hundred thousand addition operations per second[1]. (In contrast, the mobile phone in your pocket costs a few hundred dollars at most, has gigabytes of memory, and can run millions of operations per second.)

As the volume of data held on mainframe systems surged in the 1960s, data organization itself emerged as a problem. File structures tended to be complex, diverse, and unique to each application, so that every attempt to use the data required custom analysis[2]. Programs built for one customer's file system could not run on another customer's file system.

The higgledy-piggledy nature of enterprise data created a vicious cycle driving expensive customization. Programs to extract, transform, and load (ETL) had to be written as custom code; so did reports that consumed the data. Lack of standards inhibited the market for "off-the-shelf" software; and without off-the-shelf software, executives did not see value in standardization.

Much of the early work in statistics and machine learning took place in academic settings. A few large companies in certain industries used statistical methods on a large scale, but there are virtually no examples of machine learning in commercial use in this period.

---

[1]http://www.phonearena.com/news/A-modern-smartphone-or-a-vintage-supercomputer-which-is-more-powerful_id57149
[2]http://www-03.ibm.com/ibm/history/ibm100/us/en/icons/reldb/

The best example of disruptive analytics in the pre-warehouse period is credit scoring. This innovation used conservative statistical techniques and took place outside of the mainstream of enterprise information technology.

Birth of the Relational Database

In 1970, IBM Research's E.F. Codd published[3] a paper that defined the relational model of data. The relational model represents all data as tuples, or finite ordered lists of elements grouped into relations. Codd sought to define an abstraction layer for queries that would protect the user from needing to know the internal organization of the data and how it is stored. By separating the logical and physical structure of the data, it would be possible for a database administrator to optimize storage to reflect query traffic and data growth.

Codd coined the term *relational database* to characterize databases organized with a relational model. IBM started to develop a prototype database ("System R") in 1974. The project proceeded[4] through three phases:

- Development of the SQL interface (1974-1975)
- Design and build of a functioning system (1976-1977)
- Evaluation in actual use (1978-1979)

Papers published by the System R team inspired a team working at the University of California, Berkeley (UCB) to start a competing effort. The UCB team, led by Michael Stonebraker, incorporated the relational database into an existing project called Interactive Graphics Retrieval System, or Ingres. Distributed under an open source license, the original version of Ingres used a query language called Quel.

In 1977, Larry Ellison and two partners founded a company called Software Development Laboratories (subsequently renamed Relational Software in 1979, Oracle Systems in 1982, and Oracle in 1995). Also inspired by Codd's paper and by System R, Ellison and his team set out to deliver a commercially licensed relational database management system (RDBMS).

They released the first version of their database, branded as Oracle, making it the first commercially available RDBMS. Ellison had hoped to make Oracle compatible with System R, but IBM refused to release detailed information about its product. Oracle ran on minicomputers from Digital Equipment instead of IBM mainframes.

---

[3]http://dl.acm.org/citation.cfm?doid=362384.362685
[4]http://www.cs.berkeley.edu/~brewer/cs262/SystemR.pdf

IBM proceeded to develop commercial versions of System/R at a more leisurely pace. The company released SQL/DS for the VSE and VM/CMS mainframe operating systems in 1981. Two years later, IBM released DB2 for the MVS mainframe operating system.

Database architects designed early relational databases to support transactional applications, the "lowest hanging fruit" for potential investment. The idea of an analytic datastore expressly designed to support analytics is the product of a later era.

Early Business Intelligence

Before organizations had data warehouses, they met needs for management information through siloed systems, and through report-writing systems.

Larger and more sophisticated firms had Decision Support Systems (DSS) or Executive Information Systems (EIS), which were introduced in the 1960s. These systems integrated the business intelligence process from source data to user, using a proprietary file system and reporting engine. Often custom built, they were expensive to build, maintain, and modify; simply adding a report, for example, required a development project, and could take months.

In most organizations, the DSS/EIS was an extension of the financial and management accounting system, providing top managers with a more detailed view of company financials than was necessary for financial accounting. Frequently, the "Data Processing" department responsible for these applications reported to the Chief Financial Officer (CFO) or Controller.

For information needs outside of the DSS/EIS, programmers in the "DP" department wrote custom programs in languages such as RPG ("Report Program Generator"), introduced by IBM in 1959. In many cases, programs accessed data directly from source systems, such as Payroll or Accounts Payable; in other cases, the "DP" organization created periodic "snapshot" files from the source systems to support report writing.

If a report needed data from more than one source system, the report developer built the necessary consolidations into the report program itself. Ensuring consistency across reports from different systems was difficult to do in a large organization. Departments defined measures in different ways, which led to conflicting reports at the most senior levels.

By today's standards, the reporting cadence was leisurely. Most reports were published once a month. In rare cases, with the most valuable operational data, a report might be updated overnight for distribution the next day. Developing reports took time and it was expensive; due to the cost, reports were largely limited to high-level metrics that top managers needed to see on a regular basis.

Early Statistics and Machine Learning

Machine learning has roots in techniques developed by statisticians and social scientists beginning in the early 19th century. Without modern computing, however, scientists could use techniques like correlation and regression to analyze at most a few hundred cases. The introduction of statistical software packages in the late 1960s and 1970s made it possible to perform statistical analysis on larger data sets, although these tools were available only to those with access to a mainframe computer.

In 1966, Anthony Barr and one of his graduate students at North Carolina State University (NCSU), Jim Goodnight, started work on a statistical package for agricultural research[5]. The National Institutes of Health and an academic consortium, University Statisticians of the Southern Experiment Stations, provided the seed money for development. Their main goal was to reprogram existing statistical libraries for the IBM System/360 mainframe computer.

By 1970, Goodnight and Barr had a working version of the software, called Statistical Analysis System, or SAS. NCSU distributed the software for free to other universities in the consortium, and issued commercial licenses to large pharmaceutical firms and insurers with an interest in running statistics at mainframe scale[6].

Six years later, more than 100 organizations used SAS. In 1976, Goodnight and some partners acquired rights to the software and formed an independent company (SAS Institute) to develop, support, and market the product. SAS and IBM formed a close partnership; IBM needed SAS and other independent software vendors to create useful applications for its mainframes, and SAS depended on IBM's sales and marketing strength, and the imprimatur of IBM's brand.

In parallel to the development of statistical software, researchers from different disciplines worked on more general methods for machine learning. Techniques such as decision trees stem from[7] the social sciences and the need to analyze "wide" sets of categorical survey data. CHAID (Chi-Square Automatic Interaction Detection) is one of the earliest tree-building techniques implemented in software. In its most widely used form, the method dates to a publication by Gordon V. Kass in 1980 and draws on other methods developed in the 1950s and 1960s.

---

[5]http://www.biostat.wustl.edu/~phil/stuff/si.html
[6]http://www.forbes.com/2007/11/08/sas-corestates-goognight-biz-cz_rl_1108sas.html
[7]https://www.cs.nyu.edu/~roweis/csc2515-2006/readings/morgan_sonquist63.pdf

Another early decision tree technique, Classification and Regression Trees (CART), is the name of an application marketed by Salford Systems based on a paper of the same name by Leo Breiman. CART is a non-parametric algorithm that learns and validates decision tree models. Salford released its first version of CART in 1983.

In the 1940s, neuroscientists sought to understand learning by developing analog models of the brain; converted from analog models to software, this strain of research developed into what we now call artificial neural networks[8]. Early efforts showed little commercial promise. Before the 1980s researchers could not solve the so-called *exclusive-or* problem, a logic problem that can be expressed in ordinary language as "one or the other but not both". Neural networks presented with this problem failed to solve it correctly. Moreover, the algorithms that could train a neural network required computing power that was simply unavailable to most researchers.

Disruptive Analytics: Credit Scoring

One of the best examples of disruptive analytics belongs to the pre-data warehouse era. In 1956, William Fair and Earl Isaac left the Stanford Research Institute and founded Fair, Isaac, and Co., now known simply as FICO. They set out to develop computer programs that could predict behavior, which they pitched to major lenders[9].

Consumer credit in this period was quite different than it was today. Few credit cards were issued by large national issuers; most bank-issued MasterCard and Visa credit cards were issued by local banks as part of an overall banking relationship. Banks relied on local credit bureaus, customer relationships, and a considerable amount of human judgment in credit decisions; they were highly selective, and bank-issued credit cards were more difficult to get. Credit cards issued by retailers played a larger role in consumer credit.

FICO's first customers were national retail card issuers, such as Montgomery Ward. These companies had larger portfolios and could benefit from the added rigor and precision of statistical modeling. Moreover, with widely distributed operations, a large workforce and diverse customer base, large retail issuers needed the consistency afforded by a standardized credit score.

---

[8]Researchers use the term "artificial neural network" to distinguish the logical model from an actual neural network, e.g., an animal brain. Since this book is not about the nervous system, we will just use the term "neural network."
[9]http://finance.yahoo.com/news/fico-became-credit-score-100000037.html

Passage of the Fair Credit Reporting Act (FCRA) in 1970 and the Equal Credit Opportunity Act (ECOA) in 1974 changed the rules for how credit bureaus could collect information and how lenders could use credit bureau information. ECOA in particular placed an affirmative burden on lenders to demonstrate that their lending practices did not discriminate against on the basis of race, sex, marital status, and other categories. Credit scoring offered lenders a neutral[10] way to evaluate credit applications.

Meanwhile, three companies—Retail Credit Company (now Equifax), TRW Information Services (now Experian), and TransUnion—had consolidated the credit bureau industry. While none of the three had a complete national database, lenders could effectively cover the United States by working with all three.

Under FCRA, if a lender acquired a consumer credit record, the lender had to either make an offer of credit or send the consumer a decline letter stating why. However, there was no such obligation of the lender never acquired the record in the first place—in other words, if the lender asked the credit bureau to provide records only for consumers meeting certain criteria. FICO worked with lenders to define actionable selection rules that maximized the number of records selected at predictable levels of risk.

Prior to 1987, FICO built custom predictive models for each lender. In 1987, FICO developed a general-purpose credit score designed to provide consistent and stable risk predictions across all three credit bureaus. (In other words, a particular score value had the same implications for risk regardless of the source of the raw credit data.) Each of the three credit bureaus implemented the FICO scoring model in its own database and charged lenders a unit price per score.

Deployment of credit scoring on a large scale, together with the consolidation of credit bureau information, created a national market in credit cards. Firms with the best credit scoring operations built huge portfolios at the expense of local and regional issuers, many of whom exited the business. While many factors contributed to this consolidation, analytics—in the form of credit scoring and prescreening—played the key role, enabling a new business model that disrupted the industry.

---

[10]By "neutral," we mean that the credit score is influenced by such things as credit history and payment history and does not expressly take the applicant's demographics into account. Of course, some demographic groups may, on average, have better or worse credit histories than other demographic groups.

The Data Warehouse Era

In Chapter One, we discussed the use of terms such as *data warehouse, enterprise data warehouse,* and *data mart.* There is some disagreement about the precise definition of these terms; in this book, we define them thus:

- An *analytic datastore* is a collection of data that is designed and implemented to support analytics. Enterprises implement datastores with relational databases, Hadoop, NoSQL databases, proprietary software or some other organized data storage system.

- *Data warehouses* and *data marts* are analytic datastores.

- Data warehouses support many subject areas; data marts support one subject area.

- Enterprise data warehouses support all of the subject areas required by an enterprise.

The introduction of the relational database in the 1970s and 1980s revolutionized enterprise data management, much the way the Ford Model T revolutionized the automotive industry. The Model T standardized automotive production, lowering costs and making autos available to a much larger market, disrupting an existing industry of expensive handcrafted cars. In a similar way, the relational database standardized data management and opened up a market in third-party software, lowered costs, and disrupted the existing practice of custom-built proprietary applications.

Enterprises did not immediately recognize the need for a distinct database architecture for analytic datastores. The idea of the dedicated analytic datastore developed over a period of some years. Database visionary Bill Inmon argued for the concept in the 1970s[11]; Barry Devlin and Paul Murphy of IBM used the term *data warehouse* in a published[12] article in 1988.

Bill Inmon published *Building the Data Warehouse,* the first book-length theoretical text detailing an integrated design philosophy for the data warehouse, in 1992. The next year, relational database progenitor Edgar F. Codd published[13] a paper outlining the differences between Online Transaction Processing (OLTP) and Online Analytical Processing (OLAP).

Meanwhile, in the late 1970s, researchers at Caltech partnered with Citibank to detail the design of a parallel relational database expressly designed for decision support. The Caltech team founded Teradata in 1979, developed a business plan, and received venture funding in 1980. In late 1983, Teradata shipped its first system to Wells Fargo.

---

[11]http://www.webcitation.org/6dhkBqptd
[12]http://ieeexplore.ieee.org/xpl/articleDetails.jsp?arnumber=5387658
[13]http://olap.com/learn-bi-olap/codds-paper/

The introduction of data warehouses fundamentally altered the analytics value chain, the process by which information passes from a system of record to an information consumer. DSS/EIS systems supported the entire process from end to end and were often incompatible with one another. Data warehouses decoupled the process into distinct parts: ETL, which transferred data from source to warehouse; the warehouse itself; and business intelligence, which completed the value chain to the consumer.

The Enterprise Data Warehouse Movement

In the 1990s, the theory of the data warehouse and the technology of scalable relational databases joined to form what we call the Enterprise Data Warehouse (EDW) movement. The EDW movement reflected the belief that organizations *ought* to invest in centralized analytic repositories as a strategic imperative. (Not surprisingly, data warehousing vendors strongly agreed.) Three key principles drove the EDW movement:

- Enterprises should consolidate all information in a single enterprise data warehouse (EDW).

- The EDW serves as the sole source of truth for the enterprise. All other repositories are "silos" and should be discouraged.

- All data from source systems should pass through a centralized data cleansing and governance process before it is available to users, who always perform analysis downstream from the data warehouse.

Guided by these principles, the organization creates an enterprise-level abstraction, or data model, that is organized by subject areas, such as marketing, finance, and operations. The data model expresses relationships among business entities such as customers, prospects, campaigns, and transactions in a consistent way across the enterprise. The organization also defines business rules and standards governing data integrity and veracity. Working from the accepted data model and business rules, the data warehousing organization maps data from source systems to the data model, and then builds processes that extract, transform, and load (ETL) the data into the warehouse.

A data warehouse that supports all of the analytic workloads for a large organization must be big, and it must be scalable so it can expand to handle growing data volumes. Teradata led the way in the 1990s, setting records for the size of the data warehouses it could deliver: from a single terabyte (TB) in 1992 to surpassing 130 TB in 1999.

Inspired by the belief in EDWs as a strategic investment, the market for data warehousing software and services grew rapidly. IBM, Oracle, and Teradata captured the lion's share of the EDW market, while Sybase, Informix, and Red Brick offered innovative alternatives. Open standards for connectivity and the common SQL language enabled the growth of an ecosystem of vendors:

- ETL vendors, including Informatica and DataStage, offered tools to build and manage the EDW.

- Business Intelligence (BI) vendors like Business Objects, Cognos, and MicroStrategy delivered reporting and query tools that enable end users to consume data held in the EDW.

EDW advocates argued that investing in a data warehouse was less expensive than maintaining multiple overlapping decision support systems. The argument was reasonable in theory, since ETL processes could be shared by multiple BI and analytics use cases. However, it's doubtful that any organization that invested in a data warehouse ever realized any net savings from doing so.

Consistency across metrics was another selling point for the EDW. Unless an organization funnels all of its data into a standardized data platform with a uniform data model and business rules, there is a risk that different applications will produce inconsistent metrics for the same business events.

Advocates also argued that EDWs built trust and confidence in the data from standardization, cleansing, and compliance with business rules. By building confidence in the data, an EDW would contribute to building a management culture based on data and metrics.

Widespread adoption of enterprise software for Enterprise Resource Planning (ERP) and Customer Relationship Management (CRM) in the 1990s accelerated the growth of data warehousing and business analytics. Enterprise software systems capture detailed information about critical business processes, and push data quality down to the point at which it is produced. EDW developers could integrate with one source system rather than many; moreover, since enterprise software systems had predefined data models, ETL could be partially prebuilt.

In the 1990s, with few exceptions, large organizations invested in data warehouse projects. Many of these projects, however, failed to live up to expectations:

- In 1997, an analysis of IT journal articles on data warehousing identified 279 successes and 100 failures[14].

---

[14]http://www.noumenal.com/marc/dwpoly.html

- A 2002 survey[15] by the Cutter Consortium, an IT analysis firm, reported a 41% failure rate for data warehousing projects.

- In 2005, consultant Gartner predicted[16] that more than 50% of data warehousing projects would have limited acceptance or would fail within two years.

Many organizations built EDWs without a tangible business case. Leveraging a line from the 1989 film *Field of Dreams*, whose central character builds a baseball diamond in a cornfield with the expectation that the ghosts of baseball greats would play a game, EDW advocates claimed that "if you build it, they will come." Their argument was that if an organization built an EDW, end users would appear even if their use cases were not well defined at project inception.

The goal of consistent metrics also proved elusive. Companies discovered that it is difficult to create an enterprise data model, as it requires consensus across prospective users. One large U.S. consumer bank implementing a strategic CRM project established a cross-functional team to define an enterprise data model. The team wrestled with the task for 15 months before reporting to management that they were hopelessly deadlocked. They were unable to agree about how to define "customer."

In short, the inconsistent and conflicting measurements seen in "siloed" reporting systems were not simply due to outdated technology. In some cases, they reflected real and persistent conflict and disagreement within the organization, for which neither an EDW nor any other technology could provide an easy resolution.

Centralized data warehousing projects moved slowly. Implementing the foundations of an EDW could take a year or more. Adding incremental subject areas rarely took less than three to six months. Even with generous budgets and skilled development teams, functional managers faced backlogs measured in years.

As the cadence of business accelerated, the slow pace of centralized EDWs created an increasing gap between the EDW ideal and the reality of what could be accomplished within available budgets. Carefully designed enterprise data models could be rendered obsolete by a merger or acquisition. A leading bank (known for data warehousing excellence) simply loaded an acquired bank's data into its warehouse with a parallel schema instead of merging the two data structures. End users seeking an enterprise view had to write SQL code to join across the two schemas, an exceptionally difficult task. This condition existed *11 years* after the acquisition.

---

[15]http://www.networkworld.com/article/2339296/software/report--data-warehouse-failures-commonplace.html
[16]http://www.gartner.com/newsroom/id/492112

Executives who could not wait for their IT organization to deliver what they needed through an EDW turned to outside providers, including consultants, Marketing Service Providers (MSPs), and Analytic Service Providers (ASPs). Functional managers also invested in their own "rogue" data marts, operating outside the scope of IT governance. Analytics leader SAS positioned and sold its ETL and data management tooling directly to functional executives. As late as 2011, SAS earned[17] almost as much revenue from software for data marts as it did from software for predictive analytics.

The OLAP ideal had always envisioned the goal of "self-service" business intelligence, in which business users access data without IT support. Visionaries tended to believe that SQL was sufficiently similar to ordinary language that any business user would be able to query the warehouse. Few EDWs ever realized this. Often, IT organizations designed database schemas for efficient data management, and not for easy navigation. As a result, only well-trained users could query the system directly. Organizations continued to maintain teams of specialists whose sole job was to create reports and queries for functional managers. Demand for reports often exceeded supply; analysis teams had long backlogs, and incoming requests could queue up for months.

Far from being the single source of truth in an enterprise, EDWs were simply one tool among many. Often, the systems EDWs were supposed to replace were never decommissioned, so the EDW simply added to the layer of databases business users could consult. Analysts continued to write custom reports using data extracted from source systems, because it was the only way to get the information they needed.

The potential value of analytic datastores was clearly established by the early 2000s. However, executives were increasingly skeptical of the vision of a *single* EDW architecture to support *all* of an enterprise's analysis.

Appliances and Columnar Datastores

As doubts grew over the value of centralized enterprise data warehouses, database architects shifted focus to applications and solutions for analytic datastores, including marketing campaign management, risk management, and industry solutions. The new focus placed a premium on rapid deployment, simplicity, and performance.

During the 1990s, when businesses invested heavily in EDWs, many database architects repurposed software designed for use in OLTP systems for data warehouses. In most cases, they relied on custom data modeling and complex tuning to support OLAP-style workloads. Teradata was an exception—that platform was explicitly designed for OLAP—but general-purpose database software from Oracle and IBM held the largest share of the data warehouse software market.

---

[17]http://www.sas.com/content/dam/SAS/en_us/doc/analystreport/idc-ba-apa-vendor-shares-excerpt-103115.pdf

Foster Hinshaw, a data warehousing veteran co-founded Netezza in 1999. He conceived[18] the term *data warehouse appliance* to describe an entirely new way to deliver analytic datastores. Like a consumer appliance, Hinshaw argued, a data warehouse appliance should be built for a specific purpose and should deliver value immediately on delivery.

To accomplish this, data warehouse appliances had to be optimized for performance on analytic workloads, such as large block reads, table scans, and complex queries. They had to be scalable, fault-tolerant and simple to install, with little or no tuning.

Netezza announced its first data warehouse appliance in 2003. The actual technologies embedded in Netezza were not unique; Teradata had delivered massively parallel (MPP) databases for years, and Tandem pioneered fault-tolerant databases. Netezza's key innovation was to combine these technologies into a single package, converged and pre-installed on hardware for immediate availability.

Explicitly targeting analytics workloads was the key to the value of a data warehouse appliance. Conventional custom-built data warehouses required complex design, configuration, and tuning precisely because they were adapted from general-purpose databases. Targeting the analytics workload enabled Netezza to design a device that could be placed into production immediately upon delivery.

Netezza radically reduced the time needed to get an analytic datastore up and running in two ways. First, it eliminated the need for complex provisioning and tuning by bundling software and hardware. Customers did not have to buy software and hardware separately, then spend time on installation and configuration. Netezza customers purchased a device, hardware with software already deployed.

Second, Netezza eliminated the need to presummarize dimensions for users. In most cases, managers need to see summary statistics, such as total sales in a group of stores, and not a list of all transactions in those stores. Conventional data warehouses satisfied this need by pre-summarizing data into physical tables at the level of aggregation required by the business user. This approach enabled the database architect to meet the need within the performance constraints of a general-purpose database.

Presummarization, however, takes time to implement, and adds overhead to the database. Moreover, it is often very difficult for managers to specify precisely *how* they want to see business facts summarized. Many, if not most of the most, important business questions require ad hoc analysis, with aggregation rules that cannot be specified in advance.

---

[18]http://www.infoworld.com/article/2681904/database/2003-infoworld-innovator--foster-d--hinshaw.html

Netezza customers did not need to presummarize data into dimensions. Instead, they would simply load data at its most granular level and summarize data as needed for individual queries. With its high-performance design, Netezza could outperform data warehouses with presummarized dimensions, and it could do so immediately, without complex requirements analysis and data modeling.

Netezza's revenue grew[19] by more than 40% each year from 2004 through its initial public offering (IPO) in 2007. Recognizing the opportunity, other startups entered the market, including Greenplum and DATAllegro in 2003, and Aster in 2005.

Columnar databases provoked[20] considerable interest in the data warehousing community in the late 2000s. In the early 1990s, startup Expressway Technologies released the first commercial columnar database, branded as Expressway 103. Scientific database vendor Sybase acquired the company in 1995 and rebranded the software as Sybase IQ.

In 2005, Michael Stonebraker of MIT and several academic colleagues introduced[21] C-Store, an open source read-optimized columnar database. With commercial backing, Stonebraker founded Vertica to deliver an optimized version as an appliance. Shortly thereafter, a group of investors acquired the intellectual property of failed appliance vendor Xprime and rebranded the company as ParAccel, with the goal of bringing a column-oriented analytic appliance to market.

Sybase and Vertica reported successes with very large databases. In 2008, Sybase IQ delivered the first petabyte-scale data warehouse, setting a world record[22]; Vertica reported rapid growth in its installed base[23].

Startup ventures offering appliances and columnar datastores challenged the leading data warehouse vendors in the first decade of this century. By the end of the decade, the leaders had assimilated the innovators, either by acquiring them or delivering their own versions of the technology.

---

[19]http://www.sec.gov/Archives/egar/data/1132484/000095013507001814/b64586s1sv1.htm

[20]http://blogs.forrester.com/james_kobielus/10-03-19-if_queries_are_king_realm_analytic_database_does_make_columnar_heir_apparent

[21]http://db.lcs.mit.edu/projects/cstore/vldb.pdf

[22]http://www.crn.com/blogs-op-ed/207801061/sybase-iq-wins-guinness-world-record.htm

[23]http://www.dbms2.com/2009/04/25/vertica-pricing-and-customer-metrics/

- Teradata responded first, offering[24] an appliance of its own in 2008. Later, in 2011, Teradata acquired Aster.

- In the same year, Oracle announced[25] its own appliance built on hardware from Sun Microsystems. Oracle acquired Sun in 2010.

- After failing to develop its own appliance, IBM acquired[26] Netezza for $1.7 billion in 2010.

- Software giant SAP acquired Sybase in 2010 for $5.1 billion.

- EMC acquired Greenplum in 2010. In 2013, EMC spun off Greenplum to Pivotal Software, which released the software assets to open source in 2015.

- In 2011, Hewlett-Packard killed its own NeoView appliance and acquired Vertica for an undisclosed price.

- Actian, a company with a large portfolio of acquired software assets, purchased ParAccel in 2013.

The wave of acquisitions between 2010 and 2013 effectively ended the disruptive threat to data warehousing industry leaders from appliances and columnar datastores. But as we shall see later in this chapter, other even greater threats were just beginning to surface at this time.

MOLAP and ROLAP

As noted early in this chapter, the introduction of data warehouses led to a new generation of business intelligence (BI) tools. Strong standards for the SQL query language and database connectivity encouraged growth of independent BI software vendors such as Business Objects, Cognos, and MicroStrategy. These BI vendors focused their efforts on improving ease of use for the end user, and they largely left database design and construction to database vendors and systems integrators.

BI vendors split between those who supported the so-called MOLAP and ROLAP architectures. Under the MOLAP (multidimensional online analytical processing) model, the application maintains a predefined set of multidimensional data summaries ("cubes") in physical tables; when the user performs analysis, the application works with the pre-summarized cubes. The application refreshes the cubes periodically to reflect new data.

---

[24]http://www.dbms2.com/2008/09/15/teradata-data-warehouse-appliance/
[25]http://flashdba.com/history-of-exadata/
[26]http://www-03.ibm.com/press/us/en/pressrelease/32514.wss#release

The *disadvantage* of MOLAP is the requirement for pre-summarization, which constrains user analysis to dimensions spelled out in advance. While BI vendors worked on making their tools easy to use, the process of creating cubes required a developer with significant training to execute. Not surprisingly, IT departments responsible for the BI system tended to have backlogs, so the time needed to have a new cube created could be measured in months.

ROLAP (relational online analytical processing) architecture, on the other hand, does not maintain cubes as physical objects. Instead, it retains metadata from the source database as well as a logical cube specification and generates cubes on demand based on user requests. Since ROLAP tools create the user views of the data on request, they produce "fresher" analysis. In contrast, MOLAP tools build the cubes in a scheduled batch process, which can be anything from hourly to quarterly to something in between.

In the 1990s, BI tools based on MOLAP held an advantage over ROLAP tools because they did not depend on database tuning and performance for good user experience. The lower latency of ROLAP versus MOLAP was not decisive because the source data warehouses tended to be updated infrequently as well.

Data warehouse appliances and columnar datastores changed this calculus. High-performance query engines decisively shifted the balance toward ROLAP tools. Working directly with granular data, these tools could generate user views of the data in seconds, even with the largest databases. As the cadence of data warehouse updates accelerated, there was a real advantage to a BI system that could deliver the most up-to-date view of the data. Increasingly, the leading BI vendors offered tools with both MOLAP and ROLAP capabilities.

BI tools in this period also increasingly separated the user interface from the facility that performs calculations after data is retrieved from the database. With browser-based UIs, the calculation facility resided on a web server, with no calculations performed in the browser-based user interface itself.

Statistics, Machine Learning, and Data Mining

From 1984 to 2000, SAS Institute's revenue grew[27] from $50 million to more than $1 billion. During this period, the company emerged as the clear leader in commercial software for advanced analytics. SAS successfully ported its software from the IBM mainframe to a multi-vendor client-server architecture. Other software vendors, notably SPSS, continued to compete; but while SPSS developed a reputation as an easy-to-use desktop tool, commercial customers viewed SAS as an "industrial-strength" tool suitable for commercial workloads.

---

[27]http://www.sas.com/en_us/company-information.html#stats

In the 1980s, machine learning emerged as a plausible alternative to statistical methods. Software developers delivered stable implementations of the CHAID and CART algorithms mentioned previously in this chapter. Neural network researchers developed the *backpropagation* technique together with the concept of the "hidden" layer. Backpropagation enabled neural networks to solve the exclusive-or logical problem in a computationally efficient way, and hidden layers enabled neural networks to outperform linear models in many cases.

Commercial supercomputers, developed and marketed by companies like Thinking Machines Corporation and Cray Research, offered viable platforms for machine learning. Using massively parallel computing on large numbers of connected machines, supercomputers provided exceptional computing power by the standards of the day. Seeking commercial applications for their machines, supercomputer companies supported research and development in machine learning. They also offered "suites" of machine learning software that ran on their machines.

Data warehousing visionaries recognized the potential for in-database machine learning as early as the 1980s. The leading data warehouse software vendors responded by building the capability into their products:

- Teradata introduced its Warehouse Miner software in 1987.
- IBM followed with Intelligent Miner for DB2 in 1991.
- Microsoft added machine learning to SQL Server in 2000.
- Oracle acquired the software assets of Thinking Machines in 1999, rebuilt it to run in Oracle Database, and released it as Oracle Data Mining in 2002.

Large massively parallel data warehouses provided the computing power needed for machine learning. In-database tools, however, were primitive compared to tools like SAS and SPSS; despite the theoretical advantages of predictive modeling inside the data warehouse, few working analysts used the capability.

While supercomputer companies struggled to find a viable commercial business model, sales of low-cost small and mid-sized computers surged, and those computers became increasingly powerful. Independent software vendors, including HNC Software, developed software to run on these machines. NeuralWare, a company founded in 1987, delivered the first commercially available software for neural networks.

Practitioners used the term "data mining" to describe the process of discovering new, valid, and useful patterns and relationships in data. The origin of the term is unknown. Academic statisticians used the closely related terms "data dredging," "data fishing," or "data snooping" as a pejorative to describe research performed without a hypothesis.

In 1994, Integral Solutions Limited, an English company, set up a Data Mining Division to develop and market a data mining software product branded as

Clementine[28]. The ISL team had a set of software modules, developed over several years of consulting, which they combined with a visual "workflow" interface. Clementine included a number of innovative capabilities drawn from the team's experience on client projects:

- Support for "wide" tables, with variable selection tools[29].

- Tools for exploring very large decision trees.

- Continuous accuracy feedback for long-running neural network training tasks.

- Copious model accuracy diagnostics.

Clementine included decision tree and neural network algorithms together with linear models, time series, clustering, and association rules. The software featured a modular and extensible design, so that additional algorithms could be plugged in to the visual interface.

In 1996, Unica Software released its Pattern Recognition Workbench (PRW). PRW featured linear models, decision trees, and neural networks together with an optimizer to help the user train and tune predictive models.

In the 1990s, predictive analytics practitioners engaged in extended debates about the relative merits of techniques from the statistics and machine learning traditions. Leo Breiman's paper[30] "Statistical Modeling: The Two Cultures," published in 2001, captures the essence of this debate.

For many users, machine learning remained something of a curiosity through the 1990s. Analysts trained in classical statistical techniques objected to what they characterized as opaqueness and lack of interpretability. For many business problems, machine learning rarely outperformed classic methods by enough to warrant the extra time and costs.

Disruptive Analytics: Fraud Detection

In 1992, HNC Software introduced[31] Falcon Fraud Manager, an application built around a decision engine. HNC targeted credit card fraud, a problem that cost credit card issuers $500 millon in 1991, almost double the losses incurred in 1990.

---

[28]https://www.cs.bham.ac.uk/research/projects/poplog/isl-docs/1999-AISBQ-TheStoryofClementine.pdf
[29]The Clementine team defined "wide" tables as "more than 100 fields." Today, data scientists routinely work with thousands or even hundreds of thousands of variables.
[30]https://projecteuclid.org/euclid.ss/1009213726
[31]http://www.kdnuggets.com/2014/03/evolution-fraud-analytics-inside-story.html

Detecting fraud is an exceptionally challenging problem for predictive analytics. Fraudulent transactions are rare: just one in a thousand transactions and four out of ten thousand accounts are fraudulent. However, each incident of fraud is expensive; moreover, recovery is rare even when perpetrators are caught and prosecuted, so it is essential to prevent a fraudulent transaction before it is authorized.

Credit card issuers collectively authorize billions of transactions each year. There are four parties to each transaction—the cardholder, merchant, the bank that receives the transaction from the merchant, and the bank that issued the credit card. Payment networks (such as MasterCard and Visa) tie the four parties together and set rules governing how much time the issuer can take to authorize the transaction.

HNC's decision engine captured data about each card transaction presented for authorization together with profile information about the cardholder and the merchant. The engine transformed the raw data into several thousand features capturing the cardholder's spending patterns, adjusted for seasonality and other factors.[32]

The predictive model used by Falcon was a feed-forward neural net trained with a modified backpropagation training algorithm. Although Falcon updated the profiles with each transaction and computed a score in real time, the model itself was static; as of 1999, HNC reviewed the model quarterly and updated it annually.

A key element of the Falcon solution was the data consortium written into each customer contract. HNC required customers to provide detailed data transactions authorized with the system. The data consortium created network effects; as more customers licensed the system, Falcon became more valuable and attractive to the remaining customers. In the year 2000, 40 of the top 50 Visa/MasterCard issuers representing 80% of worldwide card transaction volume used the system.

Falcon is a disruptive innovation because it contributed to the consolidation of the credit card industry. Without robust fraud detection, banks had to know the customers to whom they issued cards, and merchants had to know the customers presenting credit cards. Automated fraud detection technology made it possible to expand the user base for credit cards and increase credit card acceptance by merchants. Fair, Isaac (FICO) acquired HNC in 2002 in a merger valued at $810 million ($1.1 billion in 2016 dollars).

Key Trends Today

It is not yet possible to write a history of the current era in analytics. Instead, in this section, we review two key trends influencing the analytics value chain

---

[32]http://www.amazon.com/Business-Applications-Neural-Networks-State/dp/9810240899

today: the digital transformation of the economy and the explosion of data. In the remainder of the book, we cover the specific innovations that are, in part, a response to these trends.

There is a third trend that we do not discuss in detail because it is generally understood: the long march of Moore's Law. We will discuss declining costs of computing in later chapters where there is a definite impact to a particular innovation. For example, the declining cost of computing clearly plays a role in the growth of in-memory analytics. Declining hardware costs are also clearly linked to the rise of machine learning and especially Deep Learning.

Digital Transformation of Business

In the 1990s and early 2000s, many businesses invested in enterprise software for Customer Relationship Management (CRM) and Enterprise Resource Planning (ERP). Enterprise software transforms business processes from pencil and paper or siloed systems to integrated digital workflows. This conversion from analog to digital greatly expanded the volume of enterprise data, because digital processes generate significantly more data than their analog equivalents.

This increased data volume created by enterprise transformation did not disrupt the data warehousing ecosystem. While the *volume* of data increased, the data was structured; it could be handled within the existing framework of enterprise data models by adding subject areas. Moreover, enterprise systems pushed data quality and business rules down to the point of data capture, so data produced by the systems was better quality than the systems they replaced.

In functional areas such as marketing, where digital media replaced analog media, many of the processes had always been handled by advertising agencies and marketing service providers (MSPs). For most organizations, much of the data from outsourced processes remained on the premises of third-party suppliers and wasn't visible to IT organizations. On the other hand, it was *highly* visible to digital marketing agencies, web hosting companies, and digital early adopters, who had to improvise new techniques to manage the volume.

Analyzing data from digital processes required specialized expertise. For digital marketing, for example, the analyst must understand sessionization, PageRank, cookies, tagging, and other concepts unique to the new methods. Enterprises tended to rely on their digital marketing vendors to provide these services. Hence, digital analytics tended to develop separately from traditional analytics.

The digital transformation of business processes also dramatically impacted the *cadence* of business. Digital processes are much more agile than their analog predecessors. When marketers use direct (postal) mail, for example, they measure campaign lifecycles in weeks and months; a well-organized and efficient team can field a direct mail campaign and collect responses in eight to twelve weeks. For email marketing and web media, marketers measure campaign lifecycles in hours.

This acceleration of the business cadence markedly changed the requirements for analytics. Where monthly or weekly reports were sufficient, functional managers now required daily or hourly reports, posing new demands for the entire analytics process. The accelerating cadence, plus the need to act quickly on real-time insights, encouraged managers to turn to streaming analytics, which we discuss in Chapter Six.

Functional managers, unwilling to wait for the EDW to meet their needs for analytics, bypassed the IT bottleneck, either by outsourcing analytics or by developing departmental solutions. In 1996, Gartner had confidently predicted that enterprise data warehouses would put marketing service providers (MSP) like Acxiom and Harte-Hanks out of business; instead, MSPs grew by double-digits. Vendors like SAS Institute capitalized on functional managers' frustration, pitching solutions that enabled them to develop their own departmental mini-warehouses.

The new breed of data warehouse appliances was a successful response to the demand for faster ways to analyze large masses of structured data. Startups like Netezza, Greenplum, Aster, and Vertica brought innovative appliances to the market and successfully sold in to organizations that had previously standardized on Oracle, IBM, or Teradata. Netezza proved that it could deliver an appliance that could deliver value within a day of delivery. Vertica's columnar database proved well-suited to analytic workloads at the petabyte scale.

In-memory databases, which we discuss in Chapter Five, are a natural extension of this need for speed in the analysis of structured data. Demands from functional managers for agility and immediacy place a premium on self-service analytics, which we cover in Chapter Nine.

Digital transformation of business processes fundamentally alters the "ownership" of data within the organization. In 2012, Gartner projected[33] that by 2017, the Chief Marketing Officer (CMO) would control more technology spending than the Chief Information Officer (CIO). While the magnitude of the shift is a matter of some debate[34], it's clear that as technology becomes pervasive, it is no longer possible for the CIO to lay exclusive claim to its governance.

Functional managers with P&L accountability tend to choose speed and agility over "enterprise" considerations. In other words, if one marketing solution supports rapid implementation and fast time to value, but another marketing solution is sold by the CIO's preferred vendor, the CMO will always choose the former over the latter. And, in most organizations the CMO will win that argument.

---

[33]http://www.forbes.com/sites/lisaarthur/2012/02/08/five-years-from-now-cmos-will-spend-more-on-it-than-cios-do/#2cae229d25e2
[34]http://www.cio.com/article/2975828/cio-role/as-cmos-start-to-outspend-cios-collaboration-remains-key.html

In short, the progressive digital transformation of business creates incentives for organizations to sacrifice centralized "enterprise" considerations in favor of those that produce immediate results. Cloud computing, which we discuss in Chapter Seven, supports and reinforces this movement; it enables functional managers to purchase computing resources from operating funds, which they control, rather than capital funds, which they do not.

The Data Tsunami

Unless you have been living in a cave for the past few years, it's likely that you've heard about the flood of data unleashed by our increasingly digital economy.

The first signs of this deluge emerged in 2007, when analyst firm IDC published[35] its first report on the digital universe. That report estimated a total of 161 *exabytes* of data created, captured, and replicated in the previous year, an astonishing figure in its own right. Even more stunning: IDC projected that new data would grow at the rate of 57% annually through 2010, doubling roughly every 18 months.

This new data was not transactional data. The total volume of transactions in the economy grows slowly, in the high single digits at most; and most of the transactions in the economy were already captured in the data warehousing ecosystem long before 2007.

IDC attributed expansion of the data universe to three major analog to digital conversions:

- Film to digital image capture
- Analog to digital voice
- Analog to digital TV

Of these, digital images, from snapshots on mobile phones to medical images, comprised the largest component of the digital flood.

IDC's analysis focused on the *sources* of data; the other factor influencing the growth of stored data was the radical decline in the cost of storage. Both factors played a role. If storage is expensive, organizations simply discard data or reduce the volume produced.

Storage costs per terabyte collapsed in the first decade of this century, declining by 90% from 2000 to 2005, and by another 90% from 2005 to 2010. Cheap storage means it is often cheaper and more cost effective to save all data rather than taking the time to sift through it and figure out what is valuable.

---

[35]International Data Corporation: The Expanding Data Universe: A Forecast of Worldwide Information Growth through 2010 http://www.tobb.org.tr/BilgiHizmetleri/Documents/Raporlar/Expanding_Digital_Universe_IDC_WhitePaper_022507.pdf

As it turns out, IDC's 2007 report *understated* the amount of data created and retained. The company revised its forecast upward in each of the following years, reporting[36] in 2011 that the digital universe had cracked the zettabyte barrier. Rapid growth and adoption of information-generating technologies, such as smart meters, vehicle telemetry, RFID, and intelligent sensors—what we now call the Internet of Things (IoT)—as well as accelerating social media interactions drove the expansion.

"Every two days we create five exabytes of data, or as much information as we created from the dawn of man through 2003[37]," said Google CEO Eric Schmidt, speaking at the Techonomy conference in August, 2010. He aimed too low. IDC stopped trying to report annual increments to the digital universe; in 2012, the company forecast[38] a digital universe reaching 40,000 exabytes in 2020—a figure that it revised[39] upwards by 10% in 2014.

This enormous expansion of the digital universe—from 130 exabytes in 2005 to 44 zettabytes in 2020—consisted largely of data from sources that did not exist when data warehousing visionaries codified its basic principles. Most of this data is "unstructured"—difficult or impossible to map into a relational data model: text, documents, logs, images, audio, and video. Machine learning and Deep Learning, which we review in Chapter Eight, play a critical role in understanding and interpreting this new data.

While the costs of storage hardware have collapsed, the flood of data affects many other costs, including software and personnel. The need to lower the cost of computing is one among many factors encouraging organizations to adopt open source software, which we discuss in Chapter Three. Complex and diverse data types drive the adoption of Hadoop and its ecosystem, discussed in Chapter Four.

Summary

What do we learn from a review of the history of modern analytics?

Statistics, machine learning, and data mining technologies developed separately from data warehousing and business intelligence technology. While data warehousing theorists argued that data mining "belonged" in the data warehouse, and leading database vendors delivered "in-database" data mining, actual users disagreed with the theorists. On the whole, they preferred separate tools based on servers or desktops, which had much richer functionality than the "in-database" data mining tools.

---

[36]http://www.emc.com/collateral/analyst-reports/idc-digital-universe-are-you-ready.pdf

[37]http://techcrunch.com/2010/08/04/schmidt-data/

[38]http://www.emc.com/leadership/digital-universe/2012iview/executive-summary-a-universe-of.htm

[39]http://www.emc.com/leadership/digital-universe/2014iview/index.htm

In this history, we can identify two clear examples of disruptive innovation based on predictive analytics: credit scoring and fraud detection. Tellingly, both applications also relied on data integration: for credit scoring, assembling large national databases of credit records; for fraud detection, the data sharing consortium at the heart of HNC Falcon. In both cases, the data was *external* to the customers who purchased the applications.

Innovative data warehousing *technology*, however, did not play a key role in these disruptive applications. FICO, credit bureaus, and HNC all relied on conservative and proprietary data platforms when first deployed. (These organizations eventually adopted modern data warehousing technologies, but the initial innovations used conventional file systems.)

Data warehouses lowered the cost of business intelligence and broadened access to enterprise data. That said, it is difficult to identify a single example of disruptive analytics based purely on an enterprise data warehouse. Despite the claims of their advocates, enterprise data warehouses were *not* strategic investments with disruptive potential.

The vision of the enterprise data warehouse as the single source of truth was never realized, even at the peak of the hype cycle. No large organization ever successfully integrated *all* of its data into a single centralized repository. If the vision of a unified enterprise data warehouse was unattainable in the 1990s, it is certainly unattainable today. There is simply too much data, the data is too diverse, and it moves too fast.

The enterprise data warehouse *idea* may be dead, but the data warehouses themselves will survive, like cathedrals in modern cities. Organizations that built them will not decommission them, and some may build anew. Every organization needs to measure its performance, and data warehouses are very good at providing broad access to consistent transactional metrics. But performance measurement isn't disruptive, and it isn't strategic. It is simply a cost of doing business, like pencils and office space. Consequently, the executives who manage data warehouses will be under constant pressure to deliver metrics at the lowest possible cost.

From a technical perspective, the data warehouse is itself being delayered into modular components. SQL engines, which we discuss in Chapter Four, operate independently of storage. Virtualization and cloud computing, covered in Chapter Seven, separate compute and storage from physical computing infrastructure.

Meanwhile, the most interesting, strategic, and disruptive analytics will be built outside the scope of the data warehouse, as we shall see in the chapters that follow.

Open Source Analytics

The Disruptive Power of Free

In every software category, free and open source software is a growing presence. In this chapter, we address the following questions:

- What is free and open source software?
- How can free and open source software make business sense?
- What are the leading free and open source software projects for analytics?

The most obvious attribute of free and open source software is something it lacks: software licensing fees. We will show, in this chapter, that free and open source software is viable, and that it is disrupting incumbents in the analytic software industry.

Open Source Fundamentals

The precise definition of open source software is a matter of some debate. We review the competing definitions first, then cover the fundamentals of open source software projects, including governance, licensing, code management and distribution, and the question of donated software.

T. W. Dinsmore, *Disruptive Analytics*, DOI 10.1007/978-1-4842-1311-7_3

Definitions

Words like "free" and "open" may seem unambiguous. In respect to software, however, they have multiple meanings.

Under a standard commercial software license model, the developer offers a license to use the software in return for a license fee. The license may be perpetual, or it may be limited to a specific term. There may be other restrictions as well: named users, specific computing devices, or specific applications, for example. The developer distributes the software in compiled form, often with a license key that prevents usage outside the limits of the software license. These measures protect the developer's economic interest in the software intellectual property.

Free and open source software operates under a completely different model. A developer distributes the software source code itself together with a license granting rights to examine, modify, and redistribute the code. The developer asserts no economic claim to the software intellectual property and foregoes a license fee for use of the software.

Two organizations, the Open Source Initiative (OSI) and the Free Software Foundation (FSF), define "free software" and "open source software" in slightly different ways:

- The Free Software Foundation publishes the *Free Software Definition*, which defines software to be free if the user can run, study, redistribute, and distribute modified versions of the software.

- The Open Source Initiative publishes the *Open Source Definition*, which details ten criteria for open source software, including access to source code, right to redistribute, and so forth.

The Free Software Definition defines a set of rights; if the software user is able to exercise those rights, the software is "free". ("Free" in the sense of "liberated" and not simply "at no cost".) The Open Source Definition defines a set of characteristics; if the software has those characteristics, it is "open source". Neither the Free Software Definition nor the Open Source Definition explicitly states that developers may not charge license fees; the distribution of source code makes it impossible to do so.

Suppose that Mary releases source code for some software under an open source license. Joe takes Mary's source code and redistributes it unmodified at a price of $100. Customers will soon figure out that they do not need to pay Joe for something they can get from Mary for free. If Joe conceals the source of the software by issuing it under a commercial license, Mary can sue Joe for infringing on her open source license.

The differences between the two organizations are primarily philosophical. The Free Software Foundation stems from the Free Software Movement, which argues that proprietary software and intellectual property rights in general "ought" not to exist, a perspective that is inherently political. Founded in 1985, the Free Software Foundation sponsors the GNU Project, a mass collaboration project.

In contrast to the Free Software Foundation, the Open Source Initiative takes the view that open source is a better and more practical way to develop software, and eschews political radicalism. Political differences aside, there are few practical differences between the two, and most open source licenses meet the criteria of both organizations.

While all open source software is "free" in the sense that anyone can acquire the source code without paying a license fee, not all "free" software is open source. Commercial software vendors can and do offer closed source, or proprietary software at no charge to the user. They may do this as a means to build awareness and trial of a commercial software product, to permit evaluations, or under a dual licensing model, which we discuss further later in this chapter.

In a similar vein, not all software labeled as "open" is "open source". Commercial vendors sometimes label software as "open" because it has a published API, or because it implements an open standard like ANSI SQL. Pricing and documentation do not make a closed product open; software is free and open if and only if it is distributed under a free and open source license.

While respecting the differences between the Free Software Foundation and the Open Source Initiative, in this chapter and throughout the book we will use the term "open source" to mean software that complies with the definitions of both organizations.

Project Governance

Open source software projects, like any other project, require an organization structure with clear accountabilities. They also need a legal framework to take ownership of software assets and issue licenses to users. Larger projects, such as R and Python, have their own governance framework; we discuss these separately later in this chapter. In this section, we review two entities, the Apache Software Foundation and the Eclipse Foundation, which together account for many widely used open source projects.

The Apache Software Foundation (ASF) is a 501(c)(3) charitable organization funded by individual and corporate donations, whose mission is to provide software for the public good. To support this mission, ASF provides a legal framework for intellectual property, accepting donated and contributed software and distributing software under a free and open source license. ASF currently supports more than 350 open source projects. In 2013, the latest year for which its IRS filings are available, ASF reported donations of $1.1 million and operating expenses of $653,000.

Each ASF project operates under a Project Management Committee (PMC), whose membership is elected from among the committers to the project. The PMC oversees the project, defines release strategy, sets community and technical direction, and manages individual releases. PMCs are responsible to ensure that the project follows core requirements set by ASF, such as rules governing legal aspects and branding.

Apache projects include contributors and committers. Contributors support the project in various ways, but only committers can create new revisions in the source repository.

Apache projects usually start as Incubator projects. During the Incubator phase, the project establishes its PMC, ensures compliance with Apache legal standards, and begins to build a community. When a project meets a defined set of project goals, it graduates to Apache top-level status. As of April 2016, there are 56 projects in Incubator status; of these, 20 have been in the program for more than a year. Over the life of the Incubator program, 159 projects graduated to top-level status through April 2016; 42 projects retired before graduating, mostly due to inactivity.

The Eclipse Foundation is a 501(c)(6) not-for-profit member supported corporation that supports and maintains Eclipse, an open source software development platform. The foundation also supports BIRT, a business intelligence platform discussed later in this chapter.

Open Source Licenses

Prior to implementation of the Berne Convention in 1988, software distributed without a copyright notice passed into the public domain. Under the Berne Convention, copyright attaches to software automatically when it is created. The Berne Convention also defined long time periods, or terms for copyrights; since 1988, there are no known examples of software that have reached the end of copyright and passed into the public domain. Software already in the public domain in 1988, such as the BASIC programming language, remains in the public domain.

Since copyright attaches automatically to software, open source licenses are needed to explicitly waive the copyright privilege for the user. The Open Source Initiative (OSI) and the Free Software Foundation (FSF) issue separate guidelines for open source and free software licenses, respectively.

According to Black Duck Software, a privately held company that tracks open source software projects, there are more than 2,400 unique software licenses currently in use. The most popular[1] licenses are the MIT license, GNU General Public License (GPL) 2.0 and 3.0, the Apache License 2.0, and the BSD License 2.0. All of these licenses meet both the OSI and FSF criteria.

MIT License: A free and permissive software license developed by the Massachusetts Institute of Technology (MIT). The MIT license explicitly grants the end user rights to use, copy, modify, merge, publish, distribute, sublicense, and sell the software to which it is attached.

A permissive software license grants the licensee the right to redistribute derived software under a different license. In other words, a developer can modify software obtained under a permissive license and redistribute it under a commercial license.

GNU General Public License: A free and non-permissive, or "copyleft" software license originally developed for the Free Software Foundation. GPL grants the licensee rights to run, study, redistribute, and improve the software to which it is attached. Version 2.0, released in 1991, and Version 3.0, released in 2007, differ in respect to detailed aspects of intellectual property law.

Copyleft, or restrictive software, licenses mandate that any software derived from software distributed under the license must be distributed under the same license. In other words, if a developer modifies software distributed under the GPL license, the modified software must also be distributed under the GPL license.

Apache Software License: A free and permissive software license developed by the Apache Software Foundation (ASF). All ASF projects distribute software under this license, and so do many other non-ASF projects. The license grants the user rights to use, distribute, modify, and redistribute derived software. While the user can distribute modified software under a different license, unmodified parts of the code must remain under the Apache license.

BSD License: A family of free and permissive software licenses developed in 1989 for the Berkeley Software Distribution, an operating system. In several different versions (known as the 2-clause, 3-clause, and 4-clause licenses), advocates for the BSD license argue that it is more compatible with proprietary licenses. The original BSD license does not meet OSI standards, but the modified versions do.

---

[1]http://www.blackducksoftware.com/top-20-open-source-licenses

Code Management and Distribution

Every open source project must decide how and where it should store source code, how to maintain versions and revisions, and how to distribute the code to prospective users.

Larger projects operate their own distribution platforms. The R Project, for example, operates the Comprehensive R Archive Network (CRAN), a network of more than 100 FTP and web servers around the world that store identical up-to-date versions of code and documentation for R. Similarly, the Python Software Foundation hosts a repository containing a reference implementation of Python. Commercial open source vendors generally operate their own distribution platforms.

Software projects in the early stages of development cannot afford to develop their own systems for code management. Two open source projects serve as key enablers to the open source community: Subversion and Git.

Apache Subversion is a software versioning and revision control system widely used by the open source software community and corporate users alike. Collabnet, a privately held software and application lifecycle code management company, developed the original version of Subversion in 2000 and donated it to the Apache Software Foundation (ASF) in 2009. ASF distributes the software under an Apache license. While Subversion itself lacks a user interface, numerous commercial and open source clients, or Integrated Development Environments (IDEs), work with Subversion.

Git is a free and open source distributed version control system developed beginning in 2005 by a team working on the Linux kernel; the software is available under a GPL license. Git itself operates from a command-line interface. GitHub, a privately held startup founded in 2008, offers a web-based version of Git, together with other features such as access control, bug tracking, feature requests, and task management. GitHub claims more than 12 million users and 35 million projects.

Donated Software

Few open source projects begin from scratch. In most cases, projects start with a software core developed either as an academic research project or as a commercial project. The copyright owners donate the source code to an entity committed to open source software, such as the Apache Software Foundation (ASF).

Examples of such donations include:

- Spark, a distributed in-memory project, donated to the ASF by the University of California in 2013.

- Storm, a streaming analytics engine, donated[2] to the ASF by Twitter in 2011.

- Hawq, an SQL-on-Hadoop engine, donated[3] to ASF by Pivotal Software in 2015.

- Impala, an analytic database, donated[4] to ASF by Cloudera in 2015.

- SystemML, a machine learning language, donated[5] to ASF by IBM in 2015.

The motivations behind such donations vary. In some cases, the original software developers deem the project unlikely to succeed as a commercial venture. Rather than investing further, the original developer donates the project to open source, takes a tax deduction for development costs, and harvests some goodwill.[6]

Donations may also be motivated by a change in strategic direction. Pivotal, for example, donated most of its software assets to open source when it changed its business model from software development to services delivery. In the two years prior to this change, the company earned substantially more from services revenue than it did from software licensing.

Corporate acquisitions can also trigger software donations when the acquiring company concludes that the software is not part of its core business. Twitter acquired the Storm software assets when it acquired Backtype in 2011; the company donated the software to open source soon thereafter.

In the case of academic software projects, universities rarely have either the desire or the infrastructure in place to manage software projects outside of the university's core mission. Developers at the University of

---

[2]https://blog.twitter.com/2011/a-storm-is-coming-more-details-and-plans-for-release
[3]https://blog.pivotal.io/big-data-pivotal/news/the-way-to-hadoop-native-sql?utm_source=social&account_id&utm_medium=TWITTER&PivotalBigData&utm_campaign=Products&20150930
[4]http://www.cloudera.com/about-cloudera/press-center/press-releases/2015-11-17-cloudera-proposes-to-donate-impala-and-kudu-to-the-apache-software-foundation.html
[5]http://www.theinquirer.net/inquirer/news/2413132/ibm-donates-machine-learning-tech-to-apache-spark-open-source-community
[6]Corporations may deduct the actual cost of software donated to charitable organizations. However, the Apache Software Foundation reported no non-cash contributions in 2013, the last year for which its IRS return is available.

California—Berkeley's AMPLab developed Spark as a research project; the University held the software copyright, but donated it to ASF.

A company's decision to donate a software project may or may not convey information about the project's potential value. Apache Hive, software for data warehousing on Hadoop, is a highly successful project. Facebook developed the original software and donated[7] it to ASF in 2008; since then, many other contributors have enhanced it. Hortonworks has invested in the Stinger project to improve the software, which is widely used and included in every commercial Hadoop distribution.

Hadoop is an open source framework for distributed computing and storage. We discuss Hadoop and its ecosystem in Chapter Four.

On the other hand, the bones of dead donated projects litter the open source world. Apache SAMOA, for example, a stream processing framework donated[8] to ASF by Yahoo in 2013, remains in Incubator status today, with just seven contributors over the lifetime of the project.

The Business of Open Source

Under a proprietary licensing model, the software developer invests time and money to develop software with the expectation that future revenue from software licensing fees will recoup the investment and return a profit. Once developed, the marginal cost to deliver a copy of the software to a new customer is minimal; hence successful software products deliver extraordinary returns on investment.

Copyright laws grant the developer exclusive rights to reproduce the software; hence, the developer has an economic monopoly in that product. The developer seeks to maximize the value of these rights by positioning the product to deliver unique benefits to the customer, for which there are no substitutes. This differentiation may be through software features, documentation, technical support, training, or through branding and marketing.

The customer's initial selection of the software takes place in a competitive market. While the developer tries to position the software uniquely, customers usually have several options from which they can choose and bargain for the best possible price.

---

[7]https://issues.apache.org/jira/browse/HADOOP-3601
[8]http://www.datanami.com/2013/11/25/yahoo_unveils_samoa_to_mine_multiple_data_streams/

Enterprise software, however, is complex, challenging to implement, and may require significant customization; thus, an organization that chooses to standardize on a proprietary software product risks *vendor lock-in*, a situation where the customer lacks bargaining power due to the high costs of switching.

Perpetual licenses, where the developer grants the customer a permanent right to use the software, is one response to the lock-in problem. Under a perpetual license, the customer pays the developer a one-time fee during the initial software selection, when the customer has the most bargaining power; in return, the developer grants a license to use the software "forever".

There are two issues with perpetual licenses. First, the high initial cost leads to very long evaluation and purchase cycles, and correspondingly low rates of adoption. The second issue relates to ongoing software development and maintenance. There are few cases where a software product, once developed, remains state of the art for very long. Customers expect software to improve continuously; this requires an ongoing partnership with the software developer. Rather than aligning the interests of developers and users, perpetual licenses create an incentive for software vendors to ignore customer needs once the product is sold.

Aside from the slower rate of adoption, proprietary software licensing creates three other issues that inhibit innovation.

First, there is the need for license keys and other mechanisms necessary to protect the developer's economic interest in the software by preventing unauthorized (e.g., unlicensed) use. These are not only annoying to the user, they can make it difficult to integrate software into an enterprise.

A second issue is the need to distribute the software in compiled form, so the user cannot inspect the source code. This is a critical limitation in analytics, where users rely on developers to implement algorithms accurately, and where small coding errors can produce spurious results.

Third, the proprietary licensing relationship places the developer and the user in an arms-length and even adversarial relationship. If the user is locked in to the software and switching is expensive, the developer has little or no incentive to add a requested feature. Prospective customers, who are not locked in yet, have much more leverage than existing customers when the developer sets priorities for enhancements.

The most obvious attribute of open source software is the absence of software licensing fees. While this is a clear benefit to the user, it raises an obvious question: without the possibility of earning revenue from license fees, why would anyone create software in the first place?

The motivations are mixed. In some cases, altruism and politics are clear motivators. Software developers may be inspired by a sense of purpose or a desire to create a better world. At least *some* contributions to open source projects are so motivated.

Alternatively, the developer may believe that if the software project succeeds and secures a high level of adoption, there will be ancillary opportunities to generate revenue through customization, services, training, and so forth. We discuss these approaches later in this chapter.

Without license fees, there are no barriers to trial, adoption, and use. Prospective users for open source software need not endure an extended and adversarial negotiation where they often lack the information they need to bargain in their own best interest. Instead, prospective users simply inspect the software and conduct a trial. If they are concerned about ensuring that the internal math is correct, they simply inspect the code.

With a higher rate of adoption and use than commercially licensed software, open source developers benefit from rapid and copious user feedback. Moreover, since users can donate enhancements back to the project, open source software has a potentially much larger pool of contributors. Due to the combination of these two factors, open source software projects tend to develop much more rapidly than proprietary software projects.

Open source languages are the best choice for custom development, for several reasons. First, the open and freely available source code enables enterprises to readily integrate analytic applications with other production applications. Second, the business model for open source software ensures that the enterprise fully realizes the value created by its investment in custom development.

Most commercial analytic software packages include proprietary programming languages; the SAS Programming Language (SPL) is an example. While courts have declared that the SAS Programming Language is in the public domain, code written in SPL requires a commercial runtime compiler to execute, which the user must license from SAS or a third party. This makes it a poor choice for developers and lacks the disruptive potential of open source languages.

Community Open Source

Under a community development model, a project has a broad base of contributors, who work independently with minimal guidance from a central authority. Many users are also contributors. The project itself has standards that govern code submission, as well as test protocols that limit the ability of bad actors to submit malicious code.

Community projects tend to choose one of two organizing models[9]:

- Under a *top-down* model, the project distributes source code with each release, but only a core group of developers can access and modify code between releases.

- Under a *bottom-up* model, the source code is accessible to all developers at all times.

Under a compromise model, the project has two tiers: a core platform and packages that run on the core platform. The project's governing body exercises tight control over submissions to the core platform, but minimal direct control over package submissions. A core team assumes responsibility for enhancements to the core platform, while package developers take responsibility for the packages they publish.

Commercial ventures operate on the periphery of a community open source project, offering support, consulting, education, and training. These ventures may or may not operate under the sanction of the project's governing body. They tend to have relatively little influence on the overall direction of the project.

Experienced software engineer and open source theorist Eric S. Raymond advocates for the bottoms-up model in *The Cathedral and the Bazaar*, arguing that more eyes on the software speed development, improve quality, and expedite problem resolution. On the other hand, absence of a strong central authority to establish design standards, for example, leads to complicated and inconsistent approaches to solving similar problems, making the end product difficult to navigate and use.

The best examples of community open source in analytics are the R Project and Python, which we discuss in depth later in this chapter. While Apache Hadoop is a community open source project, most organizations use commercially supported products based on Apache Hadoop, which are quite different from the open source core.

Commercial and Hybrid Open Source

Under purely commercial open source models, a commercial venture seeks to define a sustainable business model while operating within the constraints of an open source software licensing model. The venture controls project governance and most contributors are employees of the same venture. There are two distinct types of commercial open source models: the open core model and the services model.

---

[9]For more details on the two models, see *The Cathedral and the Bazaar*, available at http://www.catb.org/~esr/writings/cathedral-bazaar/.

Firms that operate under an open core model offer multiple software editions with at least one edition available under an open source license and at least one edition available under a commercial license. Typically, the commercial version has additional features that are not present in the open source edition. This model enables prospective customers to evaluate the basic software at no charge and without restriction, and provides a revenue stream from customers who choose the more feature-rich edition.

Examples of this model include:

- Talend offers Talend Open Studio as open source software and several other commercially licensed software products.

- Oracle offers the open source Oracle R Distribution, and the commercially licensed Oracle R Enterprise, which includes additional features.

Commercial ventures operating under a services model distribute software exclusively under open source licenses and sell services to users. Services may consist of cloud services, technical support subscriptions, training, and professional consulting services for implementation or custom development.

Examples of this model include:

- H2O.ai distributes H2O, an open source machine learning software package, and it sells subscription services into the user base[10].

- Google distributes TensorFlow, an open source Deep Learning software package, and it sells a managed service for the software on its Cloud Platform.

Under a hybrid business model, the commercial venture does not control project governance, but exerts strong influence over it through roles on the project's governing body. Employees of the commercial venture make important contributions to the open source project, but so do others.

The best examples of this are the Apache projects, where the Apache Software Foundation's governance model prohibits exclusive control by a commercial venture:

- Databricks leads development of the Apache Spark and offers cloud services, training, certification, and conferences.

- Hortonworks exercises strong influence over the direction of the Apache Hadoop project, and it offers its own open source Hadoop distribution together with services and training.

---

[10]As of August 2016, H2O.ai is currently testing a new product (branded as "Steam") which will be commercially licensed.

Overall, commercial ventures play a key role in open source software by promoting interest in the project and providing enterprises with the services necessary to drive value. However, there are numerous examples of widely used open source projects without a commercial ecosystem.

Open Source Analytics

Open source business models are pervasive among emerging analytics technologies. Consequently, we cover specific open source projects in other chapters:

- Chapter Four: Apache Hadoop and its ecosystem.

- Chapter Five: Apache Spark and other in-memory platforms.

- Chapter Six: Streaming analytics, including Apache Flink, Apache Storm, and other packages.

- Chapter Eight: Machine learning and Deep Learning, including CNTK, DL4J, H2O, TensorFlow, Theano, and other packages.

Among relational databases, open source MySQL and PostgreSQL ranked[11] second and fifth, respectively, in the DB-Engines Ranking in April 2016. Open source NoSQL databases MongoDB, Cassandra, and Redis ranked fourth, eighth, and ninth. Overall, open source databases account for five of the top ten most popular databases.

DB-Engines measures[12] database popularity by trac]king the number of mentions on Google and Bing, search interest in Google Trends, frequency of technical discussions on Stack Overflow and DBA Stack Exchange, job offers on Indeed and Simply Hired, profile mentions in LinkedIn, and Twitter mentions.

In January 2013, open source databases accounted for **36%** of the total popularity measured by DB-Engines; in April 2016, they captured **45%** of total measured database popularity. In certain database categories, including wide column stores, graph databases, document stores, time series databases, key-value datastores, and search engines, open source dominates.

---

[11]http://db-engines.com/en/ranking
[12]http://db-engines.com/en/ranking_definition

Three commercially driven open source projects offer integrated platforms for business intelligence: Jaspersoft, Pentaho, and Talend. All three operate under an open core model and offer commercially licensed editions with additional features.

- Jaspersoft Community includes tools for ETL and reporting (including BI for mobile devices and OLAP).

- Pentaho Community offers tools for business analytics, data integration, reporting, aggregation, schema definition, and metadata management.

- Talend Open Studio includes capabilities for scalable ETL, data quality, and master data management.

The Business Intelligence and Reporting Tools project, or BIRT, a community open source project, is a top-level project of the Eclipse Foundation. BIRT's functionality includes a report designer and report execution engine. Actuate, a subsidiary of OpenText, provides technical support and consulting, but BIRT is independently governed.

There are two commercially-driven open source projects for advanced analytics: RapidMiner and KNIME. Both projects started in Europe as academic projects and both offer visual interfaces for the business user. RapidMiner commercially licenses its most current software version and distributes prior releases under a free and open source license. KNIME distributes its base platform under open source license and offers extensions under a commercial license. We discuss these projects separately in Chapter Nine on self-service analytics.

In this chapter, we cover three analytic programming languages: R, Python, and Scala. R is a tool developed by statisticians and analysts expressly for analysis; Python is a general-purpose programming language with rich analytics capability. Scala is an elegant programming tool most notable for its strong Spark APIs.

While R's analytic functionality exceeds what is currently available in Python, Python is catching up quickly. At present, more people use R than Python for analytics, but that is also changing rapidly.

Licensing is a key differentiator between R and Python. R's GPU license is a "poison pill" for commercial developers, as products derived from R can only be redistributed as open source software. This is not an issue for Python, as the Python license is permissive. Python's governance model is also more open and broad-based, in contrast to R's closed governance.

For analysts whose primary goal is insight, R's breadth of analytic tools and visualization capabilities make it the preferred choice. For analytic developers, on the other hand, whose goal is to build applications with embedded analytics, Python's general-purpose functionality makes it the best choice.

While there is growing interest in Scala among developers and data scientists with a software engineering background, its native analytics capabilities are still limited. We include Scala in this chapter primarily because it is one of the principal interfaces to the Spark platform for distributed analytics.

The R Project

The R Project ("R") is a popular free open source language for statistical analysis, graphics, and advanced analytics. It runs on a variety of platforms and supports a wealth of functionality. While R has limits, it is so widely used by researchers and statisticians that many call it the *lingua franca* of advanced analytics.

Ross Ihaka and Robert Gentleman wrote the original code base for R in 1993, using the syntax of the S programming language. In 1995, they published the source code under the Free Software Foundation's GNU license[13].

User interest grew steadily as contributors ported existing packages to R and developed new features from scratch. Ihaka and Gentleman established the R Core Development Team in 1997 to lead ongoing enhancement to the core software environment. The R Core Development Team develops the R core software, while individual developers contribute packages with specific features. The code base is diverse; as of Release 2.13.1, 22% of the code is in R itself, 52% is in C, and 26% is in Fortran.[14]

In 2002, Ihaka and Gentleman donated the software to the R Foundation for Statistical Computing, a not-for-profit public interest organization located in Vienna, Austria. The foundation holds and administers the copyright and serves as an official voice for the project. Governance of the foundation rests in a self-selecting body of Ordinary Members, who are selected for their (non-monetary) contributions to the project; as of this writing, there are 29 Ordinary Members from the United States, Canada, United Kingdom, Norway, Denmark, Germany, France, Switzerland, Italy, Austria, India, New Zealand, and Australia.[15]

The R Foundation distributes R under the GNU license. This license makes it difficult for commercial software developers to add value to R, as any modifications or enhancements become part of the free distribution. Developers who distribute commercial applications built with R must distribute the enhanced source code. This requirement does not apply to enterprises who build applications with R for internal use only.

---

[13]https://www.stat.auckland.ac.nz/~ihaka/downloads/Interface98.pdf
[14]http://librestats.com/2011/08/27/how-much-of-r-is-written-in-r/
[15]http://www.r-project.org/foundation/

R supports analytic projects from beginning to end, including:

- Data import and export, including interfaces with most commercial and open source databases
- Custom programming, including conditionals, loops, and recursive operations
- Data management and storage
- Array and matrix manipulation
- Exploratory analysis, discovery, and statistics
- Graphics and visualization
- Statistical modeling and machine learning
- Content analytics

The R distribution includes 14 "base" packages that support basic statistic, graphics, and valuable utilities. Users may selectively add packages from CRAN or other archives. Due to the broad developer community and low barriers to contribution, the breadth of functionality available in R far exceeds that of commercial analytic software.

As of April 2016, there are 11,531 R packages in all major repositories worldwide, of which 8,239 are in the Comprehensive R Archive Network (CRAN), the most widely used R archive[16]. While these statistics demonstrate the astonishing breadth of capability included in R, they are misleading measures of usefulness. Since package developers work independently of one another, there is a great deal of overlap in functionality; a search on one repository for packages that support "linear regression" returns more than 50 packages. Package quality, documentation, and developer support is also uneven; hence, ordinary users tend to rely on a limited number of packages.

Open source R operates exclusively in memory. Lacking a capability for out-of-memory operations, R will fail if the user attempts to work with a data set that is larger than memory.

The *plyr* package provides the user with a framework to split large data sets, apply a function for each subset of data, and combine the results. The *dplyr* package extends this framework and provides a set of efficient data-handling tools together with interfaces to popular open source databases (such as PostgreSQL, MySQL, and Google BigQuery).

---

[16]http://r4stats.com/articles/popularity/

There are several other open source packages available in R for working with Big Data. These include:

- Programming with Big Data in R (*pbdR*), a suite of packages designed to support a variety of methods for high-performance computing

- Simple Network of Workstations (*snow*) supports parallel computing on a network of workstations using R

- *Rdsm* implements a threads programming environment for R, either across clustered machines or on a single multicore machine

As a rule, these packages work best for embarrassingly parallel tasks. However, many tasks in predictive analytics are not embarrassingly parallel; for these tasks, a distributed platform is the better tool. We cover these later in the chapter.

Most Big Data platforms support an R interface so that an R user can pass commands to the data platform. These interfaces do not "make R run in the database"; they convert R commands to platform-specific commands and then return the result to the user as an R object. The quality, utility, and level of support for these interfaces varies considerably across platforms; most vendors do not support or warrant the R code itself.

Microsoft offers an enhanced open source R distribution and a commercially licensed version with additional features, especially the ability to analyze data that exceeds the size of the computer's main memory. Microsoft provides technical support, training, and consulting services for organizations implementing R.

Oracle offers a free enhanced R distribution (Oracle R Distribution). It also bundles an enhanced version (Oracle R Enterprise) together with Oracle Data Mining in the Oracle Advanced Analytics Option for the Oracle Big Data Appliance. Oracle offers technical support for Oracle R Enterprise to customers who license the Advanced Analytics Option or the Big Data Appliance.

Tibco offers Tibco Enterprise Runtime for R (TERR), which is a commercial R implementation written from the ground up by professional programmers. As a result, it's generally faster and better at memory management. In particular, its ability to optimize loops written in the style of other languages is faster than open source R.

The core R distribution includes a bare-bones interface for interactive use and script development. Most users prefer to use an integrated development environment, or IDE; the most popular of these is RStudio. RStudio offers open source and commercially licensed versions of its software; the commercial version includes additional features for enterprise deployment.

It is difficult to say exactly how many people use R. In 2009, *The New York Times* reported[17] a user base of 250,000; others estimate the user base in the millions.[18]

Analyst surveys generally show R to be among the most popular tools analyst available today, but the sampling methods for these surveys make it difficult to generalize them to the population at large.

Data mining web site KDnuggets.com regularly polls its readers; each year, it asks readers to identify analytic software tools used in the past 12 months "for actual projects". In the 2016 poll, 49% of respondents said they used R more than any other tool.

Rexer Analytics conducts an annual survey of working data miners. Among analysts surveyed in the 2013 survey, 70% said they use R, up from 48% in 2011.

O'Reilly Media's tracking survey of data scientists gathers information about salaries and tool usage among working data scientists. For the 12 months prior to September 2015, 52% of respondents reported using R, which ranked third behind SQL and Microsoft Excel.

Table 3-1 summarizes results from the three surveys.

Table 3-1. Analytic Programming Tool Usage: R

			Response for R	
Survey	Date	Question	Percent Use	Rank
KDnuggets[19]	June 2016	What analytics, Big Data, data mining, or data science software, have you used (in the) past 12 months for a real project?	49%	1
Rexer[20]	Q1 2013	What (analytic) tools did you use in the past year? (Total)	70%	1
O'Reilly Media[21]	September 2015	Which programming languages do you use (for data science)?	52%	3

---

[17] http://www.nytimes.com/2009/01/07/technology/business-computing/07program.html?pagewanted=all
[18] http://blog.revolutionanalytics.com/2010/11/r-is-hot-part-5.html
[19] http://www.kdnuggets.com/2016/06/r-python-top-analytics-data-mining-data-science-software.html/2
[20] http://www.rexeranalytics.com/Data-Miner-Survey-2013-Intro.html
[21] https://www.oreilly.com/ideas/2015-data-science-salary-survey/page/4/tools-versus-tools

The TIOBE Programming Community Index is an indicator of the relative popularity of programming languages, and it combines those aimed at analytics with general-purpose languages. It is based on mind share, measured by search results and other indicators. As of May 2015, R ranks 12 out of 100 ranked languages[22], up from 33rd in May 2014. However, R is the #1 application-specific language associated with analytics (might include SAS' rank here too). (Below the top ten, rankings in this index are volatile.)

R's key strengths are its functionality, extensibility, and low cost of ownership. The free distribution eliminates barriers to entry and enables users to get started quickly.

R's key weakness is its bazaar-like nature, which appears to the novice as a plethora of conflicting and redundant functionality, loose standards, and mixed quality. While experienced users tend to revel in R's diversity and community development, users accustomed to commercial products may find it unattractive.

Python

Python is a scripting language whose syntax enables programmers to write efficient and concise code. While not as feature-rich for analytics as R, Python's capabilities for scientific computing are expanding rapidly.

In 1989, Guido van Rossum started developing Python as a hobby project while he was working for the *Centrum Wiskunde & Informatica (CWI)* in the Netherlands.[23] After using it internally at CWI, van Rossum published release 0.9.0 to alt.sources in 1991. With a growing base of users and contributors, the code base expanded steadily[24], reaching 1.0 status in 1994, 2.0 status in 2000, and 3.0 status in 2008.

In 1995, Jim Hugunin of MIT developed Numeric, a Python extension module based on ideas from the matrix-sig interest group. Over the next several years, a group of Python users from the scientific and engineering communities contributed to Numeric and developed other packages for scientific computing. In 2003, Travis Oliphant, Eric Jones, and Pearu Peterson released the SciPy package, which offered standard numerical operations running on top of Numeric. Around the same time, Fernando Perez released the first version of IPython, an interactive development environment designed to serve the scientific community.[25]

---

[22]http://www.tiobe.com/index.php/content/paperinfo/tpci/index.html
[23]http://www.artima.com/intv/pythonP.html
[24]http://python-history.blogspot.com/2009/01/brief-timeline-of-python.html
[25]http://www.computer.org/csdl/mags/cs/2011/02/mcs2011020009.html

The Python Software Foundation (PSF), founded in 2001, owns the intellectual property rights to Python Releases 2.1 and higher and issues open source licenses under the Python Software Foundation License (PSFL). PSFL is approved by the Open Source Initiative and the Free Software Foundation; it permits developers to modify the code and produce derivative works without publishing the source code. This makes Python an attractive development platform.

PSF produces the core Python distribution, which is written in C (CPython), and promotes development of the Python user and contributor community. The foundation's board consists of 11 directors, elected annually by the voting members of the community.

As a general-purpose language, Python natively supports core capabilities needed in an analytic language, such as data import and program control. For advanced analytics, two packages (*NumPy* and *SciPy*) provide foundation functions. Together, NumPy and SciPy add an interactive shell; data handling tools; multidimensional arrays, sparse matrix handling; statistical functions; linear algebra; optimization; spatial analytics; and an interface to R.

For data manipulation and analysis, many Python users work with *pandas*, a package designed to handle structured data. Pandas supports SQL-like operations, such as join, merge, group, insert, and delete. It also handles more complex operations, such as missing value treatment and time series functionality.

The richest and most widely used Python package for advanced analytics is *scikit-learn*. This package includes algorithms for classification, regression, and clustering, including logistic regression, naïve Bayes classifier, support vector machines, ensemble models gradient boosting and random forests and k-means. The package also includes tools for dimension reduction, model selection, and pre-processing.

Pybrain is designed for use by entry-level Python users. It supports a variety of techniques for both supervised and unsupervised learning, reinforcement learning, and black-box optimization. The package emphasizes network architectures.

Pattern is explicitly designed to support web mining. The package includes tools for web services, web crawling, and domain parsing, natural language processing, machine learning, and network analysis.

Many Python packages support highly specialized and advanced analytics. A few examples include:

- For anomaly detection and streaming analytics, the *NuPic* package supports Hierarchical Temporal Memory (HTM) algorithms, an extension of Bayesian techniques.
- The *Nilearn* package provides multivariate analytics for Neuroimaging data.
- *Hebel* supports GPU-accelerated Deep Learning neural networks with CUDA (through PyCUDA).

The growth rate of Python functionality is stunning. At the end of May, 2014, there were just over 44,000 packages in the Python Package Index (PyPI). A year later, at the end of May, 2015, there are just over 60,700 packages listed. Of these, 4,007 are tagged as Scientific/Engineering.

Like R, Python is memory constrained and cannot work with data sets that are larger than memory. As with R, an expert programmer can work around this constraint, but doing so negates some of the reasons for using a high-level language in the first place.

Ipython.parallel provides an architecture for parallel and distributed computing, thus enabling the user to:

- Visualize large distributed data sets with IPython
- Parallelize execution for embarrassingly parallel tasks, such as model scoring
- Write custom code to parallelize algorithms that are not embarrassingly parallel

Scalable machine learning platforms, such as Apache Spark and H2O, eliminate the need to write custom code in the latter case. These platforms support Python APIs.

Most massively parallel processing (MPP) databases run Python scripts as table functions. IBM PureData, Pivotal Greenplum, and Teradata Aster support this capability; Teradata Database does not.

Continuum Analytics publishes *Anaconda*, a free Python distribution that includes a number of enhancements for scientific computing and predictive analytics. These include:

- Pre-selected Python packages for science, math, engineering, and data analysis
- An online repository with the most up-to-date version of each package
- A set of plug-ins for Microsoft Excel

The commercial server version of Anaconda includes technical support, a private and secure package repository with a graphical user interface, customized installers and mirrors, and comprehensive licensing.

There is a wide variety of IDEs available for Python users. The *IPython* project (ipython.org) offers an architecture for interactive computing, including an interactive shell, browser-based notebook, visualization support, embeddable Python interpreters, and a framework for parallel computing.

Rodeo is a recently introduced IDE designed expressly for data science.

Python consistently ranks as one of the most popular programming languages measured by the TIOBE Programming Community Index. However, while this index tells us something about Python's overall popularity, it says little about its popularity as an analytics language. Many Python users do not use it for analytics.

In the three analyst surveys (see Table 3-2) described earlier in this chapter, Python ranks below R but above all other languages except SQL. Analysts' use of Python is growing rapidly, as is shown by the KDnuggets annual survey; in the most recent poll, reported Python use surged from 20% in 2014 to 46% in 2016.

Table 3-2. Analytic Programming Tool Usage: Python

Survey	Date	Question	Response for Python	
			Percent Use	Rank
KDnuggets	June 2016	What analytics, Big Data, data mining, or data science software, have you used (in the) past 12 months for a real project?	46%	2
Rexer[26]	Q1 2013	What (analytic) tools did you use in the past year? (Total)	24%	3
O'Reilly Media[27]	September 2015	Which programming languages do you use (for data science)?	51%	4

As a production-capable scripting language, Python is an excellent tool for analytic applications. Python supports a strong testing framework, which enables straightforward code transition from development to deployment. Since Python is widely used among developers, its use by data scientists reduces or eliminates the cultural barrier that sometimes impedes predictive model deployment.

Python's liberal open source license is another key strength. Developers may use, sell or distribute Python-based applications without permission.

Compared to R and to end user analytic tools, visualization in Python is more difficult and less compelling. While Python's statistics and machine learning

---

[26]http://www.rexeranalytics.com/Data-Miner-Survey-2013-Intro.html
[27]https://www.oreilly.com/ideas/2015-data-science-salary-survey/page/4/tools-versus-tools

capabilities are growing, it still falls short of R. Due to its history as a general-purpose scripting language, the Python community appears less attractive to business analysts and prospective users whose background is in statistics and data mining rather than software engineering.

Scala

Responding to limitations of the Java programming language, Martin Odersky started work on Scala while working at the Ecole Polytechnique Federale de Lausanne (EPFL) in Switzerland.[28] In 2004, the Scala team released the software to the public on Java and .NET. Odersky started Typesafe, Inc. in 2011 to provide commercial support and training; in 2012, the venture raised $14 million from Greylock Partners and other venture capitalists.[29]

EPFL holds the copyright to all Scala development prior to 2011; Typesafe and EPFL jointly hold the rights to enhancements from 2011 and later. End user licenses are under the open source modified BSD license.

Scala's native capabilities for analysis are immature. *ScalaNLP* is the most widely used Scala package for scientific computing and machine learning. The package includes several pertinent libraries:

- Breeze supports numerical computing, linear algebra, optimization, and signal processing

- Breeze-viz is used for visualization

- Epic is a natural language processing component with parsing capabilities for eight languages: English, Basque, French, German, Hungarian, Korean, Polish, and Swedish

- Junto supports semi-supervised learning, including label propagation, adsorption, and modified adsorption

- Nak is a machine learning library that supports k-means clustering, logistic regression, support vector machines, naïve Bayes, and neural networks

- Puck supports natural language processing on a GPU chip

---

[28]http://www.artima.com/weblogs/viewpost.jsp?thread=163733
[29]http://techcrunch.com/2012/08/22/typesafe-raises-14m-from-shasta-greylock-and-juniper-to-commercialize-scala/

Other packages for Scala include:

- Bioscala supports bioinformatics. Including DNA and RNA sequencing
- Chalk is a library for natural language processing
- Figaro is a package for developing probabilistic models

Data scientists using Scala can work with Apache Spark; interest in Spark is driving interest in Scala.

As of May, 2016, Scala ranks 32nd in the TIOBE Programming Community Index, well below R and Python. In the 2016 KDnuggets poll referenced previously in this chapter, 6% of respondents said they used Scala in the past year.

As an analytic programming language, Scala's primary strength is its Spark API, which is stronger than its Python API. Scala's main weakness is its lack of mature native analytic capabilities.

The Disruptive Power of Open Source

Open source business models disrupt established software markets in two ways.

First, in the absence of software license fees, open source software offers no initial barriers to trial. While commercial software vendors tend to develop increasingly complex and feature-rich products to justify their software license fees, open source software provides basic value at significantly lower overall cost. This is a classic example of "low-end disruption," where the functionality of existing products overshoots what many potential customers can actually use.

For example, most commercial statistical packages include a dazzling array of techniques for statistics and machine learning; they cost anywhere from hundreds to thousands of dollars. But if a practitioner simply needs to use linear regression, open source R offers an excellent alternative.

Second, due to open source software projects' rapid cadence of development and close interaction with users, developers tend to introduce the most innovative techniques in open source software first. Commercial software vendors tend to set priorities for enhancements based on short-term revenue impact; enhancements catering to niche markets or new methods that may take some time to develop tend to take a lower priority. Open source software, by contrast, offers no barriers to entry for innovators.

As an example of this process in action, consider the case of Random Forests, a machine learning technique that is highly popular today and widely used. Leo Breiman's initial paper detailing the technique first appeared[30] in 2001. Soon

---

[30]http://link.springer.com/article/10.1023%2FA%3A1010933404324

thereafter, in April, 2002, developers ported Breiman's Fortran code to R and published the randomForest package. It took another 10 years for SAS, the industry leader in commercial software for statistics and machine learning, to offer the technique in any of its products.

The inherently innovative nature of open source software development enables open source projects to provide solutions to problems that are not effectively addressed by industry incumbents. In the next chapter, we discuss open source Hadoop and its ecosystem, and how it has permanently disrupted the data warehousing industry.

The Hadoop Ecosystem

Disrupting from Below

In 2003, Doug Cutting and Mike Cafarella struggled to build a web crawler to search and index the entire Internet. They needed a way to distribute the data over multiple machines, because there was too much data for a single machine.

To keep costs low, they wanted to use inexpensive commodity hardware. That meant they would need fault-tolerant software, so if any one machine failed, the system could continue to operate.

Early in their work, they ruled out using a relational database. Their data included diverse data structures and data types, without a predefined data model. Mapping that data into a relational data structure would take too much time—if it were possible at all.

Drawing from two papers published by Google engineers, Cutting and Cafarella developed a distributed file system and programming framework. The file system, now called the Hadoop Distributed File System, or HDFS, included built-in redundancy, so that if any machine in the cluster failed, the system could continue to work with copies on other machines.

© Thomas W. Dinsmore 2016
T. W. Dinsmore, *Disruptive Analytics*, DOI 10.1007/978-1-4842-1311-7_4

The programming framework, MapReduce, provided developers with a way to distribute workloads to nodes in a cluster and collect the results. With MapReduce, the developer did not have to explicitly parallelize the task, which saved time and simplified the programming task. It also made it possible to abstract the coding layer from cluster management, so that it was possible to modify the cluster configuration without changing the program. Programs written to run on one cluster with MapReduce could run unmodified on other cluster with a different configuration without modification.

Cutting and Cafarella designed HDFS and MapReduce to work together as a system. In 2006, they contributed the code to the Apache Software Foundation as the core of Apache Hadoop.

Early versions of Hadoop were difficult to use, with few tools available to the analyst. For most enterprises, Hadoop was little more than a curiosity. Internet companies, however, adopted the technology quickly. In 2008, Yahoo announced[1] that it had set a record for a terabyte sort with Hadoop; soon thereafter, Facebook revealed[2] that it ingested 15 terabytes a day into its 2.5 petabyte data warehouse on Hadoop.

Hadoop's suitability for general use took a step forward in 2009, when Cloudera and MapR delivered commercially supported distributions. In 2013, the Hadoop team introduced YARN, a resource manager. This proved to be so significant that the community coined the term "Hadoop 2.0" to characterize the new phase of the project.

Since then, enterprise adoption has exploded:

- Wikibon estimates that one third of all organizations with Big Data have Hadoop clusters in production.[3]

- Ovum analyst Tony Baer estimates[4] that there were 1,000 Hadoop clusters implemented by early 2014.

- Leading Hadoop distributors report sales of 50 to 75 new customers per quarter.

- The most sophisticated and mature users have clusters of more than 1,000 nodes.[5]

---

[1] http://sortbenchmark.org/YahooHadoop.pdf
[2] http://blog.cloudera.com/blog/2009/05/5-common-questions-about-hadoop/
[3] http://wikibon.com/hadoop-nosql-software-and-services-market-forecast-2013-2017/
[4] http://www.enterprisetech.com/2014/10/29/hadoop-finds-place-enterprise/
[5] http://www.enterprisetech.com/2013/11/08/cluster-sizes-reveal-hadoop-maturity-curve/

- A Gartner survey[6] of its CIO Panel conducted in 2015 indicates a likely doubling of the Hadoop user base in 2016.

- Industry analyst Forrester predicts[7] that 100% of large enterprises will adopt Hadoop by 2017.

In this chapter, we cover basic principles of Hadoop and its ecosystem; the economics of Hadoop; an introduction to NoSQL datastores; and a review of analytics in Hadoop.

Hadoop and its Ecosystem

A modern relational database management system is a complex bundle of technologies: a file system, query engine, backup and recovery tools, scripting language, bulk load facilities, security systems, and administration and governance tools. Hadoop unbundles these components into many separate software components which can operate independently of the others.

There are three ways to define Hadoop:

- **Apache Hadoop**: A project of the Apache Software Foundation for distributed computing and storage.

- **"Core Hadoop"**: A set of widely used open source components that complement the functionality of Apache Hadoop.

- **The Hadoop ecosystem**: A network of open source projects, commercial software vendors, and service providers that has developed around Apache Hadoop.

Each Apache project operates independently; this enables rapid innovation and development. However, each project supports a narrow function, so that users typically combine software from multiple projects into a working system.

Commercial vendors sell bundles of open source and proprietary components, including Apache Hadoop. These are commonly called Hadoop distributions, although Apache license agreements prevent vendors from using that terminology.[8]

---

[6]http://www.gartner.com/newsroom/id/3051717
[7]http://www.networkworld.com/article/3024812/big-data-business-intelligence/the-top-5-hadoop-distributions-according-to-forrester.html
[8]http://wiki.apache.org/hadoop/Defining%20Hadoop

Apache Hadoop

Apache Hadoop is a top-level project of the Apache Software Foundation, which owns the code and distributes it under a free license to use. Hadoop operates under Apache's standard governance model. The project team consists of a Project Management Committee (PMC) and a number of committers, who are volunteer software developers. The PMC sets priorities for software enhancements and bug fixes. The team publishes a new release roughly every three months, based on a voting consensus of the committers.

Hadoop supports large-scale fault-tolerant distributed computing on commodity hardware. It includes three major components:

- **Hadoop Distributed File System (HDFS)**: A distributed file system implemented in Java.

- **MapReduce**: A processing framework for working with distributed data sets.

- **YARN**: A resource management and scheduling application.

Under Hadoop 1.0, HDFS includes a Name Node and one or more Data Nodes. The Name Node serves as a catalogue and keeps track of files stored in the system; the Data Nodes hold the data itself. Under HDFS, a file is distributed across many Data Nodes.

The MapReduce 1.0 engine consists of a Job Tracker node and one or more Task Tracker nodes. In a standard configuration, one Task Tracker and one Data Node is installed on each server in a cluster, with one or more servers designated to support the Name Node and Job Tracker.

Users query data held in HDFS by writing MapReduce programs, consisting of *mappers* and *reducers*. A mapper distributes operations such as select, filter, and sort; a reducer performs a summary operation on results of the mapper. Users can leverage these fundamental operations to support a wide range of business applications.

Prospective users can procure Apache Hadoop directly from the Apache Software Foundation.

Hadoop 2.0

In Hadoop's first generation, workload management was hardwired into MapReduce, which made it difficult to co-locate software with other processing models in the cluster. The introduction of YARN in late 2013 marks

Hadoop's second generation, or Hadoop 2.0. YARN is the result of a code rebuild that split the role of the JobTracker and TaskTracker into three separate entities:

- **ResourceManager:** A scheduler that allocates computing resources in the cluster to applications.

- **NodeManager:** Deployed on each node in the Hadoop cluster, this component manages resources on its node under the direction of ResourceManager.

- **ApplicationMaster:** Runs a specific job on the cluster, procures computing resources from ResourceManager, and works with NodeManager to manage resources available for the job.

YARN and Hadoop 2.0 had an enormous impact on business analytics. Under Hadoop 1.0, analytical applications were deployed *beside* Hadoop; users either physically extracted the data from Hadoop and moved it to the analytical application, or they converted their commands to MapReduce for direct execution in Hadoop.

Moving very large data sets is impractical (and sometimes impossible); moreover, the MapReduce learning curve is very steep even for trained analysts. A few software vendors attempted to build tools that automated the conversion to MapReduce behind the scenes, with limited success.

YARN permits deployment of analytic applications directly on the Hadoop cluster, co-located with MapReduce. The analytic applications can consume data stored in HDFS directly, without passing commands through the MapReduce engine. YARN serves as a "traffic cop" for resources, managing conflict between MapReduce jobs and the co-located software.

Powered by Hadoop

In 2009, Cloudera, a startup, launched the first commercially supported software powered by Hadoop. Cloudera bundled Apache Hadoop together with several other components to make a complete package for data management. They branded the bundle as Cloudera Data Hub (CDH), offering technical support and consulting services.

MapR, also founded in 2009, offered a competing Hadoop bundle with a number of modifications designed for high availability. MapR also included an option to substitute a proprietary file system (MapR-FS) for better performance and ease of use.

Yahoo, an early Hadoop adopter, published its own bundle for several years, but provided no technical support. In 2011, Yahoo spun off its Hadoop assets to a new company, Hortonworks, which stressed a "pure" open source approach to Hadoop. Hortonworks includes no proprietary components in its software, relying purely on technical support and consulting services for revenue.

Competition among the leading Hadoop companies is intense; each seeks to improve the value of its bundle through services, support, and product enhancements. As a result, the core Hadoop project is increasingly stable and feature-rich.

Outside of lab environments, few enterprises use pure Apache Hadoop, relying instead on one of the commercial bundles powered by Hadoop. As of July 2016, these are:

- Amazon Elastic MapReduce
- Cloudera Data Hub (CDH)
- Hortonworks Data Platform
- IBM Infosphere BigInsights
- MapR

Enterprises rely on commercial bundles for several reasons. First, commercial bundles include additional components that support capabilities needed in a data platform: implementation, maintenance, provisioning, security, and so forth.

The commercial vendors test all components in the bundle for interoperability, then provide technical support and consulting services to customers. These services provide significant value, as they simplify deployment and reduce time to value.

Commercial vendors also provide a more stable release cadence than Apache Hadoop, which releases new versions frequently and irregularly.[9] Enterprises generally prefer to avoid the cost and risk of frequent software upgrades.

Since the commercial vendors compete with one another and seek to differentiate their products, each vendor complements Apache Hadoop with a different bundle of software. The nine components listed here are included in every distribution.

- **Apache Flume**: Distributed service for collecting aggregating and moving log data
- **Apache HBase**: Distributed columnar database

---

[9]Apache Hadoop distributed 6 releases in 2011; 13 in 2012; 15 in 2013; 8 in 2014; and 5 in 2015.

- **Apache Hive**: SQL-like query engine
- **Apache Oozie**: Workflow scheduler
- **Apache Parquet**: columnar storage format.
- **Apache Pig**: High-level MapReduce scripting language
- **Apache Spark**: Distributed in-memory computing framework, with libraries for SQL, streaming, machine learning, and graph analytics
- **Apache Sqoop**: Tool for data transfer between relational databases and Hadoop
- **Apache Zookeeper**: Distributed cluster configuration service

Beyond these universally distributed components, there are many other Apache open source projects that support functions needed for data management. The projects listed next are included in some, but not all, commercial Hadoop distributions:

- **Accumulo:** Wide-column datastore
- **Atlas:** Metadata repository for data governance
- **Avro:** Remote procedure call and data serialization framework
- **Falcon:** Data governance engine that defines, schedules, and monitors data management policies
- **Knox:** System that provides authentication and secure access
- **Ranger:** Comprehensive security administration
- **Sentry:** Provides role-based authorization
- **Slider:** Tool to deploy and monitor applications running under YARN
- **Storm:** Distributed real-time computation system
- **Tez:** Software that accelerates MapReduce operations

Commercial vendors mix Apache open source projects, non-Apache open source software, and proprietary software in their products. Hortonworks bundles open source software only.

Performance Improvements

Compared to high performance data warehouse appliances, Hadoop is slow, but improving quickly. The performance issue is attributable in part to the immaturity of the platform; data warehouse appliances have sophisticated query optimizers that accelerate runtime for each operation. Optimizers for Hadoop are still in an early stage of development.

MapReduce breaks high-level tasks, such as a SQL operation, into multiple steps. It saves intermediate results to disk after each pass through the data; as a result, complex analysis runs significantly slower in MapReduce than it does in a data warehouse appliance. This is especially true for iterative algorithms, such as k-means clustering or stochastic gradient descent algorithms.

The leading commercial vendors have divergent approaches to this performance problem. Cloudera embraces Apache Spark, while Hortonworks promotes Apache Tez.

Apache Spark is a framework for distributed in-memory processing. Comparable tasks run much faster in Spark than they do in MapReduce because Spark retains intermediate results in memory rather than persisting them to disk. We cover Spark in detail in Chapter Five, which covers in-memory analytics.

Apache Tez takes a different approach. Working behind the scenes, Tez models program logic as a Directed Acyclic Graph (DAG), dynamically reconfiguring and simplifying the code. As such, Tez works in a manner similar to query optimizers in relational databases. While Tez reduces the number of MapReduce steps needed to implement a complex analysis, it does not change the need for MapReduce to persist intermediate results to disk.

Cloudera endorsed Spark in 2013 and announced that it would include Spark in its next release. The other vendors followed suit, so that by June 2014 all of the commercial Hadoop vendors had endorsed Spark and included it in their products.

In September, 2015, Cloudera announced what it calls the One Platform Initiative, under which it plans to make Spark the primary computing platform in its Hadoop release, relegating MapReduce to secondary status for new applications. This does not mean that MapReduce is dead; there is a large inventory of existing programs that will continue to run in MapReduce.

Using Tez, Hortonworks' Stinger project has improved the performance of Hive. However, planned extensions for machine learning haven't been realized, and Tez appears to be a technological cul-de-sac.[10]

---

[10]http://hortonworks.com/blog/stinger-next-enterprise-sql-hadoop-scale-apache-hive/

The Economics of Hadoop

Two key factors drive Hadoop adoption. The first of these is flexibility— the ability to capture and retain data without first defining its structure enables enterprises to act more quickly. The second is cost: Hadoop and NoSQL databases are much less expensive per unit of data than conventional data warehouses.

Hadoop and relational databases are like apples and oranges: they do very different things. Hadoop loads quickly and does not require a predefined structure; hence it is very well suited to support the flood of unstructured data produced by the new digital economy. However, Hadoop is relatively hard to query, and its runtime performance on requests is relatively slow.

Relational databases—and data warehouses based on them—require predefined data models, which can be time consuming and expensive to build. But they are relatively easy to query, and they can be tuned for extremely fast runtime performance.

Hadoop provides very low cost storage; according to one analyst, Hadoop costs[11] $1,000 to $2,000 per terabyte. In contrast to Hadoop, a Teradata data warehouse can cost[12] up to $69,000 per terabyte.

Apache Hadoop itself is free and open source; commercial bundles must be licensed, but every distributor provides a free version, enabling users to get started for little or no cost. The cost model for Hadoop scales smoothly with user needs. In other words, it is relatively easy to add another node to a Hadoop cluster, and the incremental cost of doing so is minimal.

Data warehouses from leading vendors are expensive to build, and costs are "front-end loaded"—the organization pays a steep price just to get started. Moreover, data warehouse costs tend to be "lumpy"— units of expansion are large and costly.

Hadoop disruption stems from the business model of the Hadoop ecosystem. For organizations with rapidly expanding volumes of unstructured data, Hadoop's low cost, agility, and scalability is very attractive relative to conventional data warehousing.

Data warehouses built on appliances, columnar databases, or in-memory databases remain attractive for high-value and heavily used data. For high-concurrency access to high-value data, where the organization is willing to pay a premium, high-performance data warehouses remain the best choice.

Just one of the major Hadoop distributors, Hortonworks, is a public company. Hortonworks' financial disclosures show rapid growth. In the second quarter

---

[11] http://www.statslice.com/hadoop-business-case-a-cost-effective-
queryable-data-archivestorage-platform
[12] http://blogs.teradata.com/data-points/how-illy-is-cost-per-terabyte/

of 2015, the company added 119 customers, a 27% increase over the previous quarter; revenue increased 35% over the first quarter of 2015 and 154% over the second quarter of 2014.

NoSQL Datastores

NoSQL means "not only SQL." NoSQL databases do not require the user to define a logical structure prior to loading the data; instead, the user defines structure when analyzing the data. That makes them well-suited to handling images, audio, video, machine-generated logs, documents, social media text, and other data with diverse formats.

Technically, Hadoop is a type of NoSQL datastore, and the Hadoop ecosystem includes popular NoSQL datastores like HBase. In addition, some NoSQL datastores can leverage HDFS or other components of the Hadoop ecosystem.

There are four major types of NoSQL datastore:

- **Key-value datastore:** An approach to data storage where records in a table can have varying field structures, in contrast to relational databases where each record has the same field structure. Each record has a unique key. Examples include *Redis, Memcached, DynamoDB,* and *Riak*.

- **Wide column datastore**: A hybrid of the key-value datastore and columnar datastore; highly scalable. Examples include *Apache Cassandra* and *Apache HBase*.

- **Document datastore**: Database designed for storage and retrieval of documents, such as news articles, SEC filings, or research papers. Examples include *MongoDB, CouchDB,* and *Couchbase*.

- **Graph database:** Datastore designed to represent data in the form of a mathematical *graph*, with nodes, edges, and properties. *Neo4j* is the leading example in this category.

There are many NoSQL databases currently in use. We briefly describe the five most popular[13] here:

- **MongoDB:** The most popular document database, widely used as the backend for production web sites and services. Developed in 2007 by MongoDB Inc., which provides commercial support; the software is free and open source under the GPU and Apache licenses.

---

[13]http://db-engines.com/en/ranking

- **Apache Cassandra:** A distributed wide column datastore, developed by Facebook and donated to open source in 2008. The Apache Software Foundation accepted Cassandra as an incubator project in 2009 and promoted it to top level status in 2010.

- **Redis:** A high-performance in-memory key-value datastore. Developed and supported by Redis Labs, the database is free and open source under a BSD license. Redis Labs offers Redis Cloud and Memcached Cloud as database services.

- **Apache HBase:** A high-performance NoSQL columnar datastore based on Google BigTable. In 2006 developers at Powerset used the Google BigTable framework for their own massively scalable columnar datastore to support a natural language search engine. Microsoft acquired Powerset in 2008 and donated the datastore assets to the Apache Software Foundation. At first, development proceeded as a Hadoop subproject; in 2010, Apache promoted the project to top-level status as Apache HBase.

- **Neo4j:** An open source graph database, developed and supported by Neo Technology. Neo licenses the software under GPL and AGPL open source licenses and an enhanced version under a commercial license. The first release of the product was in 2010.

At present, NoSQL datastores are used most heavily in operational applications, with limited use as analytic datastores. This is primarily due to the lack of a standard interface equivalent to SQL. Analytic programming languages, such as Python and R, can interact with the most popular databases, but use of these tools is limited to power analysts and developers.

Industry analysts expect a shakeout in the category, arguing[14] that even the most popular NoSQL databases lack a sustainable business model.

Analytics in Hadoop

Within the Hadoop ecosystem, there are a number of open source projects that support analytics: Hive, Impala, Spark, Drill, Presto, and Phoenix for SQL; Kylin for OLAP; Mahout, Spark, Flink, and H2O for machine learning; Giraph

---

[14]http://siliconangle.com/blog/2015/03/11/wikibon-view-open-source-nosql-database-vendors-face-a-long-hard-slog/

and Spark GraphX for graph analytics; Zeppelin for machine learning pipelines; and Lucene/Solr for search and text analytics. In Hadoop 2.0, there are also a growing number of commercial software packages for analytics.

We divide this chapter into two parts. In the first, we cover analytics available in Hadoop 1.0, which were rudimentary compared to what was available concurrently outside of Hadoop. In the second part, we cover analytics in Hadoop 2.0. We limit the scope of that review to SQL engines, deferring coverage of machine learning engines to Chapter Eight, and self-service BI to Chapter Nine.

Hadoop 1.0

Until the Apache Hadoop team introduced YARN in 2013, analytic workloads ran in MapReduce, because it was not possible to run alternative programming frameworks concurrently in Hadoop. Consequently, analytic tools served as brokers, translating higher-level tasks into MapReduce commands, submitting them for execution and handling the result set.

We discuss four open source projects and one commercial offering in this section. We also note that three open source analytics projects discussed elsewhere in this book—Jaspersoft[15], Pentaho,[16] and Talend[17]—pioneered integration with Hadoop.

Two additional projects are of historical interest. The rhadoop[18] project, sponsored by Revolution Analytics, supports connectivity to HDFS, HBase, and Avro, and enables dplyr-like operations in MapReduce for R users. GitHub statistics show[19] minimal activity since 2013, and virtually no activity since Microsoft acquired Revolution Analytics in 2015. The RHIPE[20] project offers comparable capability for R users. Code contributions are sporadic.

The limitations of MapReduce discussed early in this chapter impaired the performance and utility of these early efforts to bring analytics to Hadoop. Subsequently, developers have re-engineered Apache Hive to support Spark and Tez, effectively upgrading it for use under Hadoop 2.0.

---

[15]http://www.jaspersoft.com/press/jaspersoft-announces-new-hadoop-based-big-data-analytics-solution

[16]http://www.idevnews.com/stories/4429/Pentaho-Ships-BI-Analytics-Tools-for-Hadoop-Cloud

[17]http://www.infoworld.com/article/2616959/big-data/7-top-tools-for-taming-big-data.html

[18]https://github.com/RevolutionAnalytics/RHadoop/wiki

[19]https://github.com/RevolutionAnalytics/RHadoop/graphs/contributors

[20]https://github.com/saptarshiguha/RHIPE/

Apache Hive

Hive is an open source data warehouse environment designed to support SQL-like queries in Hadoop. Originally designed to run exclusively in MapReduce, Hive is now able to execute queries in Apache Tez. Hive is the most mature and most widely distributed SQL-on-Hadoop project.

Developers at Facebook started working on Hive in 2007 and donated the code to the Apache Software Foundation in 2008 as a Hadoop contributed subproject[21]. In September 2010, Hive graduated[22] to top-level Apache project status. Hive's committers include developers from technology leaders such as Cloudera, Dropbox, Facebook, Hortonworks, InMobi, Intel, LinkedIn, Microsoft, NexR, Nutanix, Qubole, and Yahoo.[23]

Prior to 2014, Hive executed queries through MapReduce. As a result, Hive's runtime performance was relatively slow, which made it best suited for Batch SQL. Under the Stinger and Stinger.next projects, Hortonworks has invested in Hive improvements, with the following primary goals:

- Improve performance to sub-second response time.
- Expand scalability to petabyte data volume.
- Enhance SQL support to full ANSI standard.

Additional enhancements planned under Stinger.next include:

- Streaming data ingestion
- Cross-geo queries, which is the ability to query data sets distributed across geographic areas
- Materialized views, which are multiple views of the data held in memory
- Ease-of-use enhancements
- Simplified deployment

For performance improvements, the Stinger team rebuilt Hive to leverage Apache Tez, an application framework that creates a more efficient execution plan than generic MapReduce. Tez models processing logic as a Directed Acyclic Graph (DAG), then dynamically reconfigures the graph for more efficient logic. According to Hortonworks, Hive on Tez runs on average 52 times faster than conventional Hive for the TPC-DS benchmark.[24]

---

[21]http://www.quora.com/How-much-time-did-it-take-to-develop-Hive-at-Facebook
[22]https://hadoop.apache.org/
[23]http://hive.apache.org/people.html
[24]http://www.slideshare.net/hortonworks/hive-on-spark-is-blazing-fast-or-is-it-final

Concurrent with the Stinger Initiative, a team at Cloudera Labs ported Hive to run on Apache Spark. The team released[25] Hive-on-Spark to General Availability in April, 2016.

Hive is distributed and supported in every commercial product powered by Hadoop, including Cloudera CDH, MapR, Hortonworks HDP, and IBM Infosphere BigInsights. A modified version of Hive is available as a cloud service in Amazon Web Services' Elastic MapReduce (EMR). This version includes the ability to import data and write back to Amazon S3 and specify an external metadata library.

Hive Server is an interface that enables users to submit queries to Hive for execution and retrieve results. (The most current version, HiveServer2, has largely displaced the original HiveServer1.) HiveServer2 supports authentication with popular security protocols, including Kerberos, SASL, LDAP, Pluggable Custom Authentication, and Pluggable Authentication Modules.

Hive supports a SQL-like language called QL ("HiveQL"). HiveQL includes a subset of ANSI SQL features, plus additional support for MapReduce, JSON, and Thrift. Users can extend Hive functionality through Java User Defined Functions (UDFs), User Defined Analytic Functions (UDAFs), and User-Defined Table Functions (UDTFs).

Hive works with data in HDFS, HBase, and compatible file systems, including Amazon S3.

According to statistics in OpenHub[26], Hive is a very active project, with 140 contributors and more than a million lines of code. The code base has expanded steadily and has accelerated markedly since 2013.

The OpenHub database, a project of Black Duck software, crawls open source code repositories to develop statistics on contributors and code commits.

Apache Pig

Apache Pig is a top-level Apache project. It includes a high-level SQL-like analytic language ("Pig Latin") coupled to a compiler that converts this language into MapReduce.

Developers at Yahoo started[27] work on Pig as a research project in the summer of 2006. Yahoo donated the software to the Apache Software Foundation,

[25]http://blog.cloudera.com/blog/2016/04/cloudera-enterprise-5-7-is-released/
[26]https://www.openhub.net/p/Hive
[27]https://developer.yahoo.com/blogs/hadoop/pig-road-efficient-high-level-language-hadoop-413.html

which accepted[28] the project for incubation in 2007. A year later, Pig graduated to top-level status.

Within the Pig team, there are ongoing projects to port Pig to Tez and to Spark.

According to OpenHub, Pig is much less active than Hive, with 28 contributors and 376 thousand lines of code. The code base peaked in 2011, with very little new activity since then.

Apache Mahout

Apache Mahout is an open source project for machine learning started[29] in 2008 by several developers from the Apache Lucene team. Initially inspired by a paper[30] authored by researchers at Stanford, the project evolved—some would say devolved—to include a mix of approaches. Instead of fostering innovation, this loss of discipline produced a loose collection of developed and contributed algorithms.

Some of Mahout's algorithms used MapReduce, others did not; some were distributed, while others ran on a single node. Many of the contributed algorithms were subsequently deprecated and removed from the project for lack of interest, use, and support.

Most of the remaining algorithms use MapReduce, but since 2014 all new algorithms must use Spark.

Beginning with version 0.11.1, which the team released in November 2015, the project includes Samsara, a math environment for linear algebra that runs on Spark or H2O.

Mahout's code base grew[31] slowly until 2012, but has not grown at all since then. It is virtually a dead project, with no code contributions in the 12 months prior to May 2016.

Apache Giraph

Graph engines perform calculations at scale on data represented as a mathematical graph. Apache Giraph is an open source iterative graph processing engine inspired by the Google Pregel graph engine described[32] in a 2010 paper; it uses a parallelization approach called *Bulk Synchronous Parallel* (BSP) processing. BSP provides a framework for managing processes and communications within a distributed processing system.

---

[28]https://developer.yahoo.com/blogs/hadoop/pig-incubation-apache-software-foundation-393.html
[29]http://www.ibm.com/developerworks/library/j-mahout/
[30]https://papers.nips.cc/paper/3150-map-reduce-for-machine-learning-on-multicore.pdf
[31]http://www.ibm.com/developerworks/library/j-mahout/
[32]http://dl.acm.org/citation.cfm?id=1807184

Giraph implements the Pregel architecture in MapReduce and works with HDFS files or Hive tables. Giraph extends the basic Pregel model with additional functionality for better performance, scalability, utility, and fault tolerance.

Facebook uses[33] Giraph to analyze the social graph of its users; the graph has more than a trillion edges. To meet requirements, the Facebook developer team modified the software and contributed the enhancements back to the open source project. The enhanced version of Giraph can:

- Read edge and node data from different data sources
- Work with any number of data sources
- Support multi-threading on Hadoop worker machines
- Make optimal use of memory on each machine
- Balance aggregation workload across machines in the Hadoop cluster

There was a surge of interest in Giraph in 2013, when Facebook first published[34] results of its assessment. Giraph's code base has grown[35] slowly but steadily since then.

Datameer

Datameer is a commercial software venture that pioneered business intelligence on Hadoop. Ajay Anand and Stefan Groschupf co-founded[36] Datameer in 2009. At the time, there were few options for analyzing data managed in Hadoop; one could write MapReduce expressions, program in Pig, or perform rudimentary SQL in an early edition of Hive. Any of these options required programming and technical skills not widely available among business analysts. Datameer set out to make Hadoop data accessible.

Datameer's product has evolved significantly since 2009, and it is now in Release 6.0 (as of May 2016). The software consists of an application server and database server that reside on an *edge node* of a Hadoop cluster.

---

[33]http://www.vldb.org/pvldb/vol8/p1804-ching.pdf
[34]https://www.facebook.com/notes/facebook-engineering/scaling-apache-giraph-to-a-trillion-edges/10151617006153920
[35]https://www.openhub.net/p/Giraph
[36]http://techcrunch.com/2010/04/13/datameer-raises-2-5-million-for-apache-hadoop-based-analytics-platform/

The Datameer application server accepts requests from end users, who work from a browser-based interface, and translates the user requests into Hadoop operations. The application evaluates the user request and submits it using one of four execution frameworks:

- **MapReduce:** The default computing framework for Hadoop, to be used if alternatives are unavailable.

- **Optimized MapReduce:** When available, Datameer uses MapReduce on Apache Tez. (For more on Tez, see the previous section on Apache Hive.)

- **Spark:** Used when Spark is available.

- **Single-Node Execution:** For small jobs, Datameer runs the request on a single node of the Hadoop cluster.

Datameer comprehensively supports major Hadoop distributions. It also supports the ability to import data from a wide variety of data sources, including relational databases, NoSQL databases, Hive, Microsoft Office formats, social media files, HTML, JSON, and many others. Data export capabilities are more limited. It is also important to note that Datameer must physically extract and move data to Hadoop for processing, as it lacks a facility for SQL pass-through to external databases.

The browser-based Datameer user interface displays results in a spreadsheet-like display, to which the user can add functions and charts.

Datameer has raised $76.5 million in five rounds of venture funding. Its most recent funding was a $40 million "D" round led by ST Telemedia, an investment firm based in Singapore. At the time of the funding announcement, Datameer claimed[37] to have 200 customers.

Hadoop 2.0

As stated previously, we limit this review to SQL engines and defer discussion of other analytic tools to later chapters.

SQL is a foundation tool in analytics, required for almost every project; many projects require nothing but SQL. SQL engines are an organic part of relational databases; organizations map data into the SQL framework when they load it into the database.

In contrast, SQL is not built into Apache Hadoop; instead, separate components called SQL engines deliver SQL capabilities. A SQL engine accepts SQL commands from the user, generates one or more requests for data, submits the requests to one or more datastores, and returns the result set to the user.

---

[37]http://techcrunch.com/2015/08/18/datameer-bags-40m-round-led-by-singapore-investment-firm/

While these engines are frequently called *SQL-on-Hadoop* engines, the term is imprecise. Some engines, like Hive and Impala, run only in Hadoop with Hadoop data sources. Others, like Spark, Drill, and Presto, can run in Hadoop or outside of Hadoop, and can work with many different data sources.

Unbundling the SQL engine from the datastore offers a number of potential benefits. The first of these is specialization: SQL engine developers can focus development on the SQL interface and query parser, while other developers enhance the datastore itself.

Separating SQL engines from the file system also enables the user to choose among multiple options. Since modern SQL engines are not commercially tied to a particular datastore, organizations can implement a "best-in-class" architecture, mixing and matching SQL engines to end user needs. Operating independently of the datastore also offers the potential to "federate" queries across multiple datastores.

Hadoop stores data without first mapping it into a SQL framework. Consequently, in Hadoop a SQL user must map data into a table structure before running queries.

Modern SQL engines support one or more of the following three modes of SQL processing:

- **Batch:** SQL scripts run without human supervision or attention on static data. Batch mode is suitable for long-running queries or queries that are scheduled for repeated execution. Typically used for Extract, Transform, and Load (ETL) processing, scoring, or scheduled reports.

- **Interactive:** SQL scripts run on static data while the user awaits a response. This mode generally requires lower latency: up to 20 minutes. Typically used for ad hoc queries and discovery, where questions may be nested.

- **Streaming:** SQL scripts run continuously on dynamic data over a sliding time window. Streaming mode requires very high performance. Typically used for algorithmic trading, real-time ad targeting, and similar applications.

Hive, Spark SQL, and Impala support HiveQL; Presto and Drill support ANSI SQL. All engines support User Defined Functions (UDFs). Hive and Spark offer query fault tolerance; the rest do not. Hive, Spark SQL, and Impala are the most mature and feature-rich; Presto and Drill are the least mature.

User Defined Functions (UDFs) are expressions, functions, or code snippets provided by a database user to supplement built-in functions. UDF support varies with each platform and can include the ability to run programs written in languages such as C, Java, Python, and R.

There are now far too many SQL engines for Hadoop to cover all of them in detail. We cover Impala, Drill, and Presto in this chapter and Spark SQL in Chapter Five. Several additional projects merit brief coverage:

- **Apache Tajo:** Tajo is an open source high-performance SQL engine. The Apache Software Foundation accepted Tajo for incubation in March 2013 and promoted it to top-level status in March, 2014. Gruter, a startup based in South Korea, leads Tajo development and offers commercial support.

- **Apache Kylin:** Kylin offers fast interactive ANSI SQL and MOLAP cube capabilities together with integration with BI tools and enterprise security. Originally developed by eBay, Kylin was accepted as an Apache Incubator project in November 2014 and graduated to top-level status in November 2015.

- **Apache Phoenix:** Phoenix is a relational database framework designed to integrate with Apache HBase; It includes a query engine, metadata repository, and JDBC driver. Phoenix converts user requests into native HBase calls, bypassing MapReduce; this enables it to run much faster than early versions of Hive. Engineers at Salesforce.com developed Phoenix for internal applications; the company donated[38] the project to open source in 2013. The Apache Software Foundation promoted[39] Phoenix to top-level status in May 2014. Hortonworks includes Phoenix in its product.

- **Apache Trafodion:** Trafodion is a SQL-on-HBase engine that offers ANSI SQL support and ODBC/JDBC connectivity for BI. HP Labs launched Trafodion as an open source project in June 2014 and released it to production in January 2015. The Apache Software Foundation accepted Trafodion as an incubator project in May 2015.

- **Apache HAWQ:** Currently in Apache Incubator status, HAWQ is a SQL on Hadoop engine evolved from Pivotal Greenplum Database. Pivotal Software donated the project to open source in June 2015.

---

[38]http://www.infoq.com/news/2013/01/Phoenix-HBase-SQL
[39]https://developer.salesforce.com/blogs/developer-relations/2014/05/apache-phoenix-small-step-big-data.html

- **IBM Big SQL:** Big SQL is IBM's SQL interface to its Hadoop distribution, InfoSphere BigInsights; it can query HDFS, HBase, and special tables created by Big SQL itself. The product includes a facility to copy data from relational databases into Big SQL tables.

- **Oracle Big Data SQL:** Big Data SQL is an Oracle product available only on the Oracle Big Data Appliance. It federates queries across Oracle Database, Oracle Hadoop, and Oracle NoSQL database in a single query. The product operates through external table extensions to Oracle Database and offers a capability called Query Franchising, through which agents on the data subsystems execute the query using equivalent operators.

Many different factors influence runtime performance of the engines. These factors include the volume and type of data, storage formats, system infrastructure, deployment characteristics, and the nature of queries tested. Not surprisingly, published benchmarks produce conflicting results due to differences in the factors cited here.

Apache Spark

Apache Spark is a distributed in-memory computing framework, with libraries for SQL, streaming, machine learning, and graph analytics. With more than 1,000 contributors, it is the most active Apache project, the most active project in the Hadoop ecosystem, and the most active project anywhere in Big Data. Given its significance, we treat Spark separately in Chapter Five, which covers in-memory analytics.

Apache Impala

Impala is a massively parallel processing (MPP) SQL platform for Hadoop. Developed and maintained by Cloudera, the software is free and open source under an Apache license. Cloudera announced[40] the project in October 2012 and released[41] it to general availability in May 2013. In 2015, Cloudera donated Impala to the Apache Software Foundation.

Apache Impala runs fast interactive SQL queries on data stored in popular Apache Hadoop file formats. Impala integrates with the Apache Hive metastore to share databases and tables between the components. This enables users to freely choose to work with Impala or Hive without moving or duplicating data.

---

[40]http://www.zdnet.com/clouderas-impala-brings-hadoop-to-sql-and-bi-7000006413/
[41]http://blog.cloudera.com/blog/2013/05/cloudera-impala-1-0-its-here-its-real-its-already-the-standard-for-sql-on-hadoop/

Impala supports HiveQL, with built-in functions for mathematics, data type conversion, date and time operations, conditional expressions, and string functions. For more advanced operations, Impala supports aggregate functions with statistics such as count, sum, mean, and median; and window functions for ordered and grouped statistics. Finally, Impala supports basic user-defined functions (UDFs) and user-defined aggregate functions (UDAFs) for custom operations

Users working with Impala submit commands through the Impala Shell or through any ODBC- or JDBC-compliant client.

Impala operates directly with the stored data, and does not use an intermediate computing layer (such as MapReduce or Spark). It works with HBase and HDFS, including common formats: text, SequenceFiles, RCFiles, Apache Avro, and Apache Parquet. Impala also works with Amazon Web Services' S3 file format.

In early 2014, IBM Research published[42] the results of a performance test that compared the performance of Hive, Hive on Tez, and Impala with different file formats and compression codecs. The test protocol used standard test protocols published by the Transaction Processing Council.

In IBM's testing, Impala ran 3-4 times faster than Hive on MapReduce and 2-3 times faster than Hive on Tez for TPC-H benchmarks. For TPC-DS benchmarks, Impala ran 8-10 times faster than Hive on MapReduce and about 4 times faster than Hive on Tez.

TPC-H and TPC-DS are standard decision support benchmarks published by the Transaction Processing Performance Council. The benchmarks consist of a series of typical ad hoc queries that simulate the work of a typical user.

Later in 2014, Cloudera published[43] results of its own benchmark testing comparing performance of Impala, Spark SQL, Facebook Presto, and Hive on Tez. For single-user queries, Impala outperformed all alternatives, running on average seven times faster. For multi-user queries, Impala widened the performance gap, running on average 13 times faster.

Cloudera, MapR, Oracle, and Amazon Web Services distribute Impala; Cloudera, MapR, and Oracle provide commercial build and installation support.

---

[42]http://www.vldb.org/pvldb/vol7/p1295-floratou.pdf
[43]http://blog.cloudera.com/blog/2014/09/new-benchmarks-for-sql-on-hadoop-impala-1-4-widens-the-performance-gap/

Apache Drill

Apache Drill is an open source distributed software framework for interactive analysis.

In 2010, a group of Google engineers published[44] a paper describing a distributed system for interactive ad hoc query analysis designed to aggregate trillion-row tables in seconds. They called the system Dremel. Dremel inspired Google's BigQuery, an interactive ad hoc query system hosted in the Google cloud.

A group of contributors led by Ted Dunning of MapR proposed to develop an open source version of Dremel, renamed Drill. In September, 2012, the Apache Software Foundation accepted Drill as an incubator project. Drill graduated to top-level project status in December 2014. In June 2015, the team published Release 1.0 of the software. Drill's developer community includes employees of MapR, Intuit, Hortonworks, Elastic, LinkedIn, Pentaho, Cisco, and the University of Wisconsin, among others.

Drill offers the user the ability to query data with or without predefined schemas, and it can query unstructured or nested data. It offers full ANSI-standard SQL, and it integrates with widely used BI tools. Drill can federate queries across multiple data sources.

Prospective users can download and install Drill directly from the project web site; MapR also distributes Drill as an add-on to its Hadoop distribution.

Users can deploy Drill as a standalone application, or on any Hadoop cluster.

End users can query Drill with BI tools (such as Tableau, MicroStrategy, or Microsoft Excel) through ODBC and JDBC drivers. Drill also supports a REST API for custom applications as well as Java and C applications.

Drill currently supports the following data sources:

- **Hadoop:** Apache Hadoop, MapR, Cloudera CDH, and Amazon EMR

- **NoSQL:** MongoDB and HBase

- **Cloud storage:** Amazon S3, Google Cloud Storage, Azure Blog Storage, and Swift

While the Apache Drill project team claims trillion-row scalability, there are no published benchmarks or reference users as of June 2015. Drill has high potential as an interactive query tool, but limited commercial adoption at present.

Drill has an active contributor base, and its code base has steadily expanded[45] since mid-2014.

---

[44]http://research.google.com/pubs/pub36632.html
[45]https://www.openhub.net/p/incubator-drill

Presto

Facebook has one of the largest active data warehouses, with more than 300 petabytes of data stored in a few large Hadoop clusters. Addressing a need for SQL connectivity to this asset, Facebook engineers created Hive software in 2007 to query this massive datastore.

As we noted previously, in its original architecture Hive used MapReduce as a compute engine, which made it suitable for batch queries only. In 2012, Facebook engineers developed a better tool for fast interactive SQL, called Presto. After extensive internal testing and rollout to the internal user community, Facebook shared the code with some other organizations for testing and feedback. In late 2013, Facebook donated the Presto code to open source and made it generally available to any user under an Apache 2.0 license.

Facebook reports that it has successfully scaled a Presto cluster to 1,000 nodes. The company also reports[46] that more than 1,000 employees run queries on Presto, and they run more than 30,000 queries per day on more than a petabyte of data. In May 2015, developers reported significant speedup in new releases through continuous improvement.[47]

Presto supports ANSI SQL queries across a range of data sources, including Hive, Cassandra, relational databases, or proprietary file systems (such as Amazon Web Services' S3). Presto queries can federate data from multiple sources. Users can submit queries from C, Java, Node.js, PHP, Python, R, and Ruby. Airpal, a web-based query execution tool developed by Airbnb, offers users the ability to submit queries to Presto through a browser.

Presto's user base currently includes Facebook, Airbnb, and Dropbox.

Organizations can deploy Presto on-premises or in the cloud through Qubole. In June, 2015, Teradata announced[48] plans to develop and support the project. Under an announced three-phase program, Teradata proposes to integrate Presto into the Hadoop ecosystem, enable operation under YARN, and enhance connectivity[49] through ODBC and JDBC.

OpenHub statistics show[50] that Presto is a very active project, with a steadily expanding code base.

---

[46]https://www.facebook.com/notes/facebook-engineering/presto-interacting-with-petabytes-of-data-at-facebook/10151786197628920
[47]https://code.facebook.com/posts/370832626374903/even-faster-data-at-the-speed-of-presto-orc/
[48]http://www.teradata.com/News-Releases/2015/Teradata-Launches-First-Enterprise-Support-for-Presto/?LangType=1033&LangSelect=true
[49]http://money.cnn.com/news/newsfeeds/articles/prnewswire/CL12937.htm
[50]https://www.openhub.net/p/facebookpresto

Summary

While there is some debate among analysts about Hadoop's growth rate and enterprise penetration, there is no doubt that the Hadoop ecosystem is growing rapidly. We know this from disclosures by Cloudera, Hortonworks, MapR, Amazon Web Services, and other companies that play a key role in the ecosystem.

There are two reasons for this growth. The first is the tsunami of text, images, audio, video, and other data that is unsuited to relational databases, as noted in Chapter Two. Economics is the second reason. Hadoop's cost per terabyte is well below the cost of alternative data management tools.

Under Hadoop 1.0 prior to 2014, Hadoop's ability to handle non-relational data was its primary justification. It was not a credible competitor to data warehouse appliances for analytics, because it was too immature, unstable, rudimentary, slow, and hard to use.

Since the implementation of Hadoop 2.0, the number of commercial and open source software packages for Hadoop has exploded. The ability to run alternative programming frameworks together with MapReduce makes it possible to run the most complex analysis.

Concurrently, the platform itself has matured. Thanks to the contributions of commercial vendors engaged with enterprise customers, Hadoop is increasingly stable and secure. Projects to improve performance, such as Tez and Spark make Hadoop a credible platform for interactive analytics. We will cover Spark in detail in Chapter Five.

For SQL analytics, Hadoop is increasingly competitive with relational databases. Under Hadoop 1.0, Hive and Pig were suitable for high-latency batch programs; semantically, they supported a subset of ANSI SQL. Thanks to the Stinger project, Hive is now competitive with other tools for interactive analysis, and so are Impala, Spark SQL, Drill, and Presto. And, they are rapidly approaching full ANSI SQL compliance.

Moreover, Hadoop is increasingly accessible to the business user. Thanks to a number of innovations in self-service analytics, it is now possible for business users to work directly with Hadoop using tools like Excel and Tableau. We cover these innovations in Chapter Nine.

As Hadoop matures, it competes with conventional data warehouses, and its attractive economics matter. Enterprises with existing warehouses have a powerful economic incentive to substitute Hadoop for existing platforms as those platforms approach end of life.

Hadoop's economics make it attractive for use cases that cannot support the costs of conventional data warehouses. More importantly, Hadoop's marginal "sunk" costs make it ideal for use cases where the returns are simply unknown. Thus, Hadoop becomes the platform of choice for labs, sandboxes, skunk works, and innovations of all kinds.

In-Memory Analytics

Satisfying the Need for Speed

"In-memory analytics" is a misnomer: all analytics run in memory and have always done so. Two things distinguish modern in-memory analytics:

- The ability to persist large amounts of data in memory, so it is immediately available for analysis, without a disk read operation.

- Scale-out capability, which is the ability to distribute large in-memory workloads over many servers.

The growth of modern in-memory analytics is partly attributable to technical innovations in database design. Database architects have developed ways to address the inherent volatility of in-memory data structures through replication, snapshotting, and other means to ensure fault tolerance and data durability.

Arguably, though, the principal force behind in-memory analytics is the declining cost of memory. While memory remains an order of magnitude more expensive than disk storage, costs have declined by more than four orders of magnitude since 1990, as shown in Figure 5-1.

© Thomas W. Dinsmore 2016
T.W. Dinsmore, *Disruptive Analytics*, DOI 10.1007/978-1-4842-1311-7_5

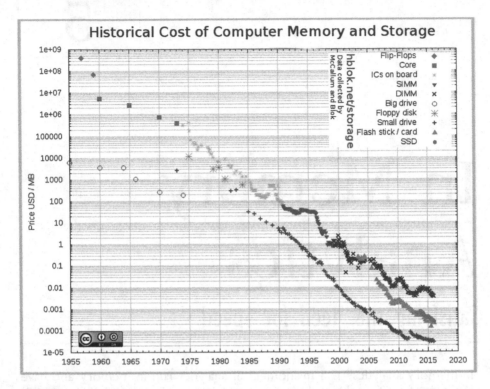

Figure 5-1. Historical cost of memory and storage (Source: McCallum and Blok, hblok. net/storage)

Since it is very unlikely that we need immediate access to all of our data all of the time, it makes good sense to design systems with more durable storage than memory.[1] This is a key issue in analytics architecture: how to balance the tradeoff between the speed of in-memory operations with the additional cost.

In the first section of this chapter, we discuss the growth of in-memory technology for relational databases. In the second section, we discuss open source memory-centric processing frameworks, including Apache Spark, memory-based file systems, and memory caching frameworks.

In-Memory Databases

In its simplest form, an analytic operation has three parts:

- Read data from durable storage.

- Perform a computation in memory with the data.

- Write results back to durable storage.

---

[1]Solid state devices and flash memory present a new opportunity for systems architecture. As of 2016, use of SSD and flash in analytic datastores is an emerging technology.

Computers perform analytic computations in random-access memory (RAM). The actual technology used for memory has changed significantly since the earliest computers, from hard-wired circuitry to single in-line memory modules (SIMMs), to the dual in-line memory modules (DIMMs) used today. But the principle is the same: all computations take place in some kind of memory.

DIMMs, like most previous memory technologies, are volatile: we lose the information in memory if the machine loses power. To avoid this loss, we save the product of the computation to persistent and durable storage that retains information without power. In the relational database era, that storage is usually a disk drive.

RAM operates much faster than the input/output (I/O) operations that move data back and forth from storage to memory. Thus the total time to complete the analytic operation depends mostly on the time needed to read and write. It takes a fraction of a second to move an item of data from disk to memory; for most conventional transaction processing workloads requiring single-record lookup, this data transfer does not create a performance bottleneck if the database is properly configured and tuned.

On the other hand, for the analysis of large data sets, the cumulative effect of this internal data movement across millions or billions of data items is significant. The problem is even more serious when the size of the data set needed for computations exceeds available memory. In some analysis software, the operation simply fails; in others, the software swaps data from memory to disk, seriously impairing performance.[2]

Columnar serialization, discussed in Chapter One, mitigates the problem by organizing the stored data in a manner that expedites its retrieval. However, it does not eliminate the I/O bottleneck, which remains measurable at terabyte and petabyte data volumes.

Caching objects in memory is one way to eliminate the bottleneck. If a task requires a series of computations on the same data, we can save time by keeping the data in memory (caching) until we reach the end of the series. At that point, we can either truncate (drop) the objects from memory to make room for the next task, or we can retain them just in case the user decides to run the task again. Either way, however, our ability to use caching depends on the amount of memory available.

While caching reduces the need for *subsequent* read operations, it does not eliminate the initial read. A more sophisticated form of caching is proactive or predictive: instead of waiting for a process to request data, it anticipates the request (based on historical usage patterns, for example) and loads the data into memory. This approach may not help the ad hoc user on every problem, but it's helpful for the most frequently used.

---

[2]http://www.phuse.eu/download.aspx?type=cms&docID=2847

Of course, if we maintain objects in a memory cache, we need to be concerned about keeping the data consistent with the disk datastore. For this, we need write-through, read-through, and write-behind capabilities. *Write-through* means the in-memory cache propagates in-memory updates to the disk datastore. *Read-through* means that if an operation requests data from the cache that it has not previously loaded, the cache retrieves the item from the disk datastore. *Write-behind* means the cache updates the disk datastore asynchronously; in other words, it completes the in-memory update for the requesting application and updates the disk datastore in the background.

A full in-memory database copies the entire database into memory, using write-through, read-through, and write-behind operations to maintain consistency with a mirrored database on disk. This mitigates the need for predictive caching, since all of the data resides in memory all of the time. Of course, as with all relational databases, the organization chooses what data to include in the database schema and what to exclude; thus, an in-memory database does not eliminate the need to organize data and set priorities.

Three key constraints limited the deployment of in-memory databases prior to 2010—cost, availability of memory, and durability.

As noted at the beginning of this chapter, the cost of memory has declined radically in absolute terms; between 2003 and 2014, the cost per terabyte of in-memory storage has declined by 95%. This opens up use cases for in-memory databases that were infeasible as recently as ten years ago.

The cost of disk storage has declined by an equal amount, and memory remains two orders of magnitude more expensive than disk storage. Thus, there is still an economic incentive to set priorities for data, maintaining only the most valuable data in memory.

Also, keep in mind that an in-memory database is actually two databases: the durable database on disk and its mirror in memory; the database architect does not choose between disk and memory, but balances the proportions of both. Combined with the extra cost of sophisticated tools to ensure durability, in-memory databases remain much more expensive per terabyte than any other mainstream data management technology.

A second constraint is the availability of memory in sufficient quantities to support large databases. In the 1990s, server vendors shipped machines with a few gigabytes of memory at most; they have increased the maximum supported memory to the terabyte range, but data volumes are expanding much faster than available memory.

To address this constraint, database architects distribute data over many servers. We discussed the general principles of distributed architecture in Chapter Four. Distributed databases date back to the 1970s, but have progressively become mainstream as engineers solve key architectural problems. Growing acceptance of Hadoop and distributed NoSQL databases reflect the increasing acceptance of scale-out architecture for large-scale data management.

The term ACID is an acronym for Atomicity, Consistency, Isolation, and Durability, a set of desirable properties for database transactions.

The third issue for in-memory databases is durability of the data: the "D" of ACID. Database designers use a number of techniques to ensure that organizations do not lose data when a database shuts down:

- Snapshots or checkpoints, which record the state of the database at a moment in time

- Transaction logging, which records changes to the database in a journal

- Non-volatile RAM, memory that retains information when powered down and can securely reproduce the state of memory on shutdown

- Replication of the data on clustered computers, with automatic failover in the case of node failure

Snapshots are only as good as the last snapshot, and any activity recorded since the last snapshot is lost. To avoid data loss, snapshots must be frequent. Growth of real-time database updates and high-velocity data renders this approach obsolete.

Transaction logging is a reasonable alternative to snapshots. However, in a database with a heavy workload, reconstructing the state of the database on shutdown can take a considerable amount of time.

The cost of non-volatile, or flash storage, has declined even more than volatile memory, to the point that it is now cheaper per terabyte than conventional (DIMM) memory. There are some disadvantages to flash memory for enterprise applications—it is less durable for frequent reads and writes—so its use is still largely experimental. In 2015, researchers at MIT demonstrated[3] the promise of this technology with a cluster of flash-based servers, claiming performance and cost effectiveness comparable to RAM-based servers.

---

[3]http://www.computerworld.com/article/2947614/cloud-storage/mit-proves-flash-is-as-fast-as-ram-and-cheaper-for-big-data.html

High availability through replication is an important byproduct of a distributed scale-out architecture. Hadoop's original developers envisioned deployment on large clusters of commodity hardware and expected a high incidence of node failures. Consequently, they built redundancy into the Hadoop Distributed File System (HDFS), with data distributed in three replicates across the cluster.

Most analytic databases work in read-only mode; in other words, users consume and analyze data but rarely create it or write back into the datastore. Hence, many data warehouses operate on a "backup is reload" principle: if the database goes down, it will be restored simply by repeating the load process that created it in the first place. Consequently, the "durability" issue that is so critical for transactional databases is less compelling for analytic datastores.

More memory capacity is now available to support in-memory databases. For scaling up on a single machine, server manufacturers now offer terabytes of memory capacity. For scaling *out*, distributing in-memory databases across multiple machines is now an option. In-memory databases are not yet able to support the largest databases, but enterprise software vendor SAP has demonstrated[4] databases of 100TB on its HANA appliance.

The leading business analytics vendors have largely assimilated in-memory technology.

Oracle, the leading data warehouse platform revenue, positions Oracle Database system as a hybrid technology, using the term System Global Area to refer to the shared-memory realm. For additional license fees, Oracle customers can license Oracle Database In-Memory, a columnar in-memory datastore closely integrated with Oracle Database. When deployed on an Oracle appliance, Oracle mirrors the data on multiple nodes for fault-tolerance and durability. Oracle also markets Coherence, an in-memory data grid acquired in 2007.

Enterprise software vendor SAP has focused its data warehousing strategy around its HANA in-memory columnar database. SAP developed HANA by consolidating numerous acquired technologies, including the TREX columnar search engine, P*TIME in-memory OLTP platform, and MaxDB in-memory caching technology. SAP's early adoption of in-memory technology, combined with its large base of existing customers, have contributed to its double-digit revenue growth in data warehousing.

Characteristically, IBM has spread its in-memory investments across many platforms and products. IBM BLU Acceleration is a bundle of technologies available for deployment with DB2, Informix, or as a service in the cloud (branded as DashDB). BLU includes a columnar in-memory datastore with data compression and hardware-based CPU acceleration.

---

[4]http://siliconangle.com/blog/2012/05/16/sap-hana-now-the-biggest-in-memory-database-in-the-world/

Microsoft SQL Server 2014 included a capability to retain entire tables in memory. Branded alternatively as Hekaton and SQL Server In-Memory OLTP, it is designed for transaction processing workloads rather than analytical ones.

Teradata does not offer a complete in-memory database. Instead, it offers what it brands as Teradata Intelligent Memory, a predictive caching capability that tracks data usage and keeps the most heavily used data in memory.

Among startups leveraging in-memory database technology, the most promising build on an open core business model and target new markets. Two startups, MemSQL and VoltDB, stand out in the so-called NewSQL category targeting Hybrid Transactional/Analytical Processing, or HTAP use cases. Both companies offer an open source version of their in-memory database and stress integration with open source Hadoop, Spark, and Kafka. These companies are betting that the future of real-time analytics rests in the open source ecosystem.

MemSQL, founded in 2011and funded by leading venture capitalists, first released its product in June 2012. The company's eponymous database software is a distributed in-memory platform that uses write-ahead logs and database snapshots to preserve data durability. For advanced analytics, the product supports geospatial functions and an interface to Apache Spark. In its 2015 analysis of in-memory database vendors, Forrester estimates[5] that MemSQL had about 50 customers.

VoltDB supports a commercial version of H-Store, an academic project led by MIT's Michael Stonebraker. Like MemSQL, VoltDB is a distributed in-memory database, with transaction logging and snapshots to ensure data durability. VoltDB 5.0, released in 2015, includes import and export integrations for Kafka and import integration with HP Vertica, as well as support for Apache Hive and Apache Pig.

Other open source in-memory databases include Aerospike, Apache Geode, Hazelcast, MonetDB, and Redis. Proprietary offerings include EXASolution from EXASOL and the Kognitio Analytical Platform from Kognitio.

Apache Spark

Apache Spark is an open source system for fast and general large-scale data processing. It provides a runtime environment for high-performance low-latency execution in several forms, including exploration, stream processing, ad hoc SQL, machine learning, and graph analytics. Spark users with a fault-tolerant and implicitly parallel interface to manipulate distributed data.

---

[5]https://www.forrester.com/The+Forrester+Wave+InMemory+Database+Platforms+Q3+2015/fulltext/-/E-res120222

The foundation of Spark is an abstract data structure called Resilient Distributed Datasets, or RDDs. RDDs are read-only partitioned collections of records distributed over a cluster of machines. Spark creates RDDs through deterministic operations on stable data or other RDDs. RDDs include information about data lineage together with instructions for data transformation and persistence. They are fault tolerant; if an operation fails it can be reconstructed.

Spark users can either retain the RDD in memory or write the results to persistent storage. This contrasts sharply with MapReduce, which requires the user to write data to storage at the end of each Reduce operation. This persistence in memory makes it possible to write iterative algorithms, query data interactively, or perform streaming operations.

More than 1,000 developers have contributed to Spark (through April 2016). With more than 30,000 commits over the lifetime of the project and more than 10,000 in 2015 alone, Spark is the most active Big Data project today.

Matei Zaharia and colleagues at AMPLab, a collaborative at University of California, launched Spark in 2009 as a research project for machine learning with large data sets. In early 2010, they released the software to open source under a BSD license. The University of California donated the software assets to the Apache Software Foundation, which accepted the project for Incubator status in June 2013. In February 2014, Spark graduated to top-level Apache status.

Databricks, a commercial venture founded in 2013, leads development of Apache Spark. In two rounds of venture funding, Databricks has raised a total of $47 million through early 2016. In the most recent round, in June 2014, a group led by New Enterprise Associates, Andreesen Horowitz, and Data Collective invested $33 million in a Series B round[6].

Spark follows the standard Apache governance model. A Project Management Committee (PMC) comprised of 35 committers (as of April 2016) oversees development of the project. Databricks employees hold 13 of the 35 seats; other entities represented include University of California, Berkeley; Cloudera; Yahoo; IBM; Intel; and eight other organizations. As of April 2016, there are 44 Spark committers, of whom 17 are Databricks employees, and 5 are affiliated with University of California, Berkeley.

Spark development follows the Apache voting process, where changes to the code are approved through consensus of committers. The team uses a review-then-commit model, where at least one committer other than the patch author reviews and approves the patch before it is merged, and any committer may vote against it.

---

[6]http://techcrunch.com/2014/06/30/databricks-snags-33m-in-series-b-and-debuts-cloud-platform-for-processing-big-data/

The PMC has designated individual committers as maintainers for certain modules to ensure consistent design for public APIs and complex components. Maintainers for impacted modules review patches before code merge.

As is the case for all Apache projects, the reference code is available directly from the Apache Software Foundation. All major Hadoop distributors include Spark in their distributions: Cloudera, Hortonworks, MapR, IBM, and Amazon Web Services. In September 2015, Cloudera announced what it calls the One Platform Initiative, under which it plans to make Spark the primary computing platform in its Hadoop release, relegating MapReduce to secondary status.

Databricks has certified additional distributions from a number of vendors, including:

- BlueData, a private cloud vendor

- DataStax, distributor of the Cassandra NoSQL datastore

- Guavus, an operational intelligence vendor

- Huawei, a telecommunications solutions vendor

- Lightbend, formerly Typesafe, developers and distributors of the Scala language

- Oracle, for the Oracle Big Data Appliance

- SAP, an enterprise software vendor

- SequoiaDB, a NoSQL datastore distributor

- Stratio, a Big Data platform vendor

- Transwarp, a Shanghai-based Big Data vendor

Spark users can deploy the software on a single machine; in a free-standing cluster; in Hadoop, running under YARN; on Apache Mesos, a distributed resource manager; on cloud platforms, including Amazon Web Services, Microsoft Azure, Google Compute, and in OpenStack; and in Docker or Kubernetes containers. For users who prefer not to provision and install the software themselves, there are a number of providers offering Spark as a managed service, including Altiscale, Amazon Web Services, Databricks, Google, IBM, Microsoft, and Qubole.

Spark has no native file system. Instead, it includes adapters that enable it to work with many data platforms:

- Hadoop Distributed File System (HDFS)

- Cloud datastores, such as AWS S3 and Redshift

- Relational databases, such as MySQL, PostgreSQL, and any JDBC-compliant RDBMS

- Common Hadoop formats, such as ORC, Parquet, and Avro files

- NoSQL datastores, such as Cassandra, Cloudant, Couchbase, HBase, MongoDB, and SequoiaDB

- Streaming sources, such as Apache Kafka

- In-memory file systems, such as Alluxio

- Search engines, such as Elasticsearch

- Connectors to applications, such as Salesforce.com and SAS

- Mainframe data

- ESRI Magellan geo-spatial libraries

- Miscellaneous data sources, such as cookie data sets, Google spreadsheets, and many others

Capabilities

The Spark Core includes foundation capabilities, including task dispatching, scheduling, and basic input/output. Working through the Spark APIs, users invoke parallel operations on RDDs by passing functions to Spark, which schedules execution in parallel on the cluster.

Each operation takes one or more RDDs as input and produces new RDDs. Spark keeps track of the lineage of RDDs, so it can reconstruct any operation if one or more machines fail. Spark uses lazy evaluation, which means that it postpones evaluation of an expression until it is needed; this improves performance and minimizes memory usage, because it avoids needless calculations.

Users interact with Spark through programming interfaces for Scala, Java, Python, and R. The Spark functions available in each API are somewhat different; the Scala API is the most highly developed, while the R API is least developed.

In addition to the Spark core for distributed data processing, the Spark project includes four libraries:

- Spark SQL, a set of tools for working with structured data

- Spark Streaming, for streaming analytics

- Spark MLLib, for machine learning

- GraphX, for graph-parallel processing

The project also includes the Spark Packages library, which includes more than 200 additional packages[7] contributed by third-party developers.

SQL Processing

Spark SQL is a component of Apache Spark that supports SQL-like processing of structured data. Respondents to a survey[8] of Spark users by startup Databricks in 2015 report that they use Spark SQL and supporting DataFrames more than any other Spark component.

In 2011, developers at the University of California at Berkeley's AMPLab began work on a SQL engine called Shark. The Shark project started with the Hive code base and replaced the execution engine with Spark's in-memory processing. This approach yielded significant improvements to query runtime; however, the team found Hive's large code base to be unwieldy and difficult to optimize.

The Spark development team introduced[9] Spark SQL as an alpha component in May 2014. In July 2014, the development team announced[10] that they would abandon further development of Shark and focus all future effort on Spark SQL, which graduated[11] from Alpha in March 2015, with Spark Release 1.3.

Spark SQL enables users to combine the concise and declarative syntax of SQL with the power of procedural programming languages. It accomplishes this through two components: the DataFrame API, which supports relational (SQL) operations, and the Catalyst optimizer, an engine that converts SQL expressions to efficient Spark operations.

DataFrames are distributed collections of structured data with named columns; they are an abstraction for selecting, filtering, aggregating, and plotting structured data. Spark users manipulate DataFrames with Spark's procedural API or with a relational (SQL) API.[12]

Spark users create DataFrames from existing Spark Resilient Distributed Datasets (RDDs), from Hive tables or directly from data sources. Spark supports native integration with Parquet data sets, JavaScript Object Notation (JSON) data sets, Hive tables, or relational databases through Java Database Connectivity (JDBC).

---

[7]234 packages as of mid-June 2016.
[8]https://databricks.com/blog/2015/09/24/spark-survey-results-2015-are-now-available.html
[9]http://spark.apache.org/releases/spark-release-1-0-0.html
[10]https://databricks.com/blog/2014/07/01/shark-spark-sql-hive-on-spark-and-the-future-of-sql-on-spark.html
[11]http://spark.apache.org/releases/spark-release-1-3-0.html
[12]http://people.csail.mit.edu/matei/papers/2015/sigmod_spark_sql.pdf

Users interact with DataFrames directly with SQL, with a programming tool, with another Spark component, or with an external BI tool. For SQL users, Spark SQL currently supports HiveQL syntax, including UDFs and UDAFs. Other Spark components, such as the machine learning library, can create and use DataFrames. BI tools like Tableau, Zoomdata, and Qlik can interact with DataFrames through a standard JDBC connector.

The Catalyst optimizer, built into the Scala programming language, converts the logic of a SQL expression into an optimal physical plan for execution in Spark. Separating the logical and physical plans enables Spark's developers and third parties to readily add new data sources as well as new language bindings. The Catalyst optimizer also ensures consistent performance across language APIs. The optimizer itself is easy to extend and enhance by adding new optimization rules and code-generation methods.

For streaming in SQL, developers at Intel contributed[13] an open source library that works with Spark SQL. This library supports time-based windowing aggregation and joins. As of late 2015, it supports the Scala interface only.

Streaming Analytics

Spark Streaming is an extension of Spark, added in 2013, that supports fault-tolerant processing of live data streams with high throughput.

Spark Streaming ingests data from streaming sources, processes the data, and pushes it to target systems. Streaming sources can include Kafka, Flume, Twitter, ZeroMQ, Kinesis, or TCP sockets, among others. Processing includes complex transformations expressed as high-level functions such as *map*, *reduce*, *join*, and *window*. Target systems can include file systems, databases, BI systems for live visualization, or other Spark libraries.

To process streaming data, Spark Streaming divides the data into microbatches, which it processes through the Spark engine. Users can define the duration of the batch window down to half a second. Spark Streaming provides high-level abstractions called DStreams or discretized streams, represented as a sequence of Resilient Distributed Datasets (RDDs), which Spark holds in memory.

Examples of Spark Streaming's transformations include:

- **Map:** Returns a new DStream by passing each element of the source DStream through a specified function.

- **Filter:** Returns a new DStream by selecting only those records of the source DStream for which a function evaluates as true.

---

[13]https://github.com/Intel-bigdata/spark-streamingsql

- **Union**: Returns a new DStream that contains the union of the elements in the source DStream and a second specified DStream.

- **Count**: Returns a new DStream by counting the number of elements in each RDD of the source DStream.

- **Reduce**: Returns a new DStream by aggregating the elements in each RDD of the source DStream using an associative function.

Transformations create new DStreams and RDDs without altering the source DStreams and RDDs. Thus, Spark can reconstruct the data with all changes in the event of a system or node failure.

Spark 2.0, released in spring 2016, embraces a new approach to streaming with Structured Streaming, which handles low latency, interactive and batch elements in a single API. Structured Streaming defines a *stream* as a high-level concept, similar to a *table* in SQL. The user simply specifies a stream as the target of an operation; behind the scenes, the Spark optimizer routes the query to the appropriate static data or data stream, as required. While a query against a finite table ends when all records are processed, a query against a stream runs until it is terminated by the user.

Spark Streaming, like other Spark libraries, supports Scala, Java, and Python APIs.

Machine Learning

Spark has two native machine learning libraries and a set of third-party libraries under Spark Packages. The two native libraries are:

- MLlib, the original API built directly on Spark RDDs
- ML, a higher-level API built on Spark DataFrames

There is some functional overlap between the two libraries. The ML API, first introduced in Spark 1.2, is easier to use and recommended for new users. While the Spark team plans to continue to support MLlib, all new algorithms will be contributed to the ML library.

We cover the details of Spark's machine learning capabilities in Chapter Eight.

Graph Analytics

Spark includes the GraphX graph engine. GraphX supports widely used graph analytics such as Page Rank, Connected Components, and Triangle Counting.

Like Apache Giraph, GraphX is inspired by the Google Pregel paper and uses the Bulk Synchronous Processing Model for graph analytics. Unlike Giraph, which is implemented in Java on MapReduce, GraphX runs on Spark. The cadence of development on GraphX is slower than it is for other Spark libraries.

Spark in Action

To illustrate how organizations adopt and use Spark, we present 10 brief examples of Spark in action.

- *Barclays*, a leading multinational banking and financial service company, uses Spark to support an "insights engine": an application that combines hundreds of queries to compute Key Performance Indicators (KPIs) for a business banking client. With 1,296 queries against 700 million records for each of 275,000 clients, Barclays could not run this analysis in its Teradata data warehouse. In Spark, the analysis runs in 30 minutes.[14]

- *BlackRock*, a leading investment and risk advisory firm, manages or advises asset and derivative portfolios valued at more than $8 trillion. Following the 2008 financial crisis, issuers disclose more detailed information about assets backed by mortgages, credit cards, and other types of consumer debt; however, this data tends to be very noisy and conventional data quality tools do not scale. Blackrock uses a Spark-based framework to create, run, and manage data quality tests on terabyte-scale data sets.[15]

- *Comcast* collects significant amounts of data about its customers, including usage clickstreams and contact events such as telesales and e-mails. The volume, variety, and velocity of this data make conventional machine learning algorithms impractical. Comcast uses Spark to detect anomalies in customer activity that may indicate service interruptions.[16]

---

[14]http://www.slideshare.net/SparkSummit/hundreds-of-queries-in-the-time-of-one-gianmario-spacagna
[15]http://www.slideshare.net/SparkSummit/topnotch-systematically-quality-controlling-big-data-by-david-durst
[16]http://www.slideshare.net/SparkSummit/petabyte-scale-anomaly-detection-using-r-spark-by-sridhar-alla-and-kiran-muglurmath

- *Goldman Sachs* has invested its core competency in analytics to build powerful data computation frameworks into its intra-company platform. Today, Goldman actively uses Spark to build data pipelines, manipulate data for reports, and tap streaming data. Working with Spark and R, users manipulate and reduce data for local analysis, plotting, and reporting, while other users run Python-based simulations in Spark.[17]

- *MediaMath*, a digital media buying service, uses Monte Carlo simulations built into Spark to test digital advertisement lift and effectiveness. Cookies are randomly assigned to test or control, and those in test are exposed to ads while those in control are not. Spark provides MediaMath with the processing power to quickly test millions of trials over tens of millions of simulated consumers.[18]

- *Netflix* used Apache Spark to build a customer simulator enabling it to "back-test" strategies and changes to its recommendation engine. The simulator allows the data science team to evaluate the impact of improvements to the algorithm, or simply to try new ideas. By using the simulator, testing places no stress on production services, and the data science team does not have to wait for history to accumulate.[19]

- *NBC Universal* stores hundreds of terabytes of media files for international cable TV distribution; efficient management of this online resource is necessary to support distribution to international clients. The company uses Spark's machine learning library to predict future demand for each item based on a combination of measures. Based on these predictions, the company moves media with low predicted demand to low-cost offline storage. The predictions from machine learning are far more effective than arbitrary rules based on single measures, such as file age. As a result, NBC Universal reduces its overall storage costs while maintaining client satisfaction.[20]

---

[17]http://www.slideshare.net/SparkSummit/how-spark-is-making-an-impact-at-goldman-sachs-by-vincent-saulys
[18]http://www.slideshare.net/SparkSummit/monte-carlo-simulations-in-adlift-measurement-using-spark-by-prasad-chalasani-and-ram-sriharsha
[19]http://www.slideshare.net/SparkSummit/distributed-time-travel-for-feature-generation-by-db-tsai-and-prasanna-padmanabhan
[20]https://spark-summit.org/2015/events/use-of-spark-mllib-for-predicting-the-offlining-of-digital-media/

- *Novartis* uses Spark to analyze data captured in assays, or laboratory experiments conducted to test hypotheses about the biology of a disease. Due to advances in biotechnology, a single screening assay used today can produce trillions of data points. Using Spark together with a visualization tool and Cassandra NoSQL datastore, Novartis performs complex analytics: normalization against controls, reduction of highly correlated features, multi-parametric classification, and more. The end result: faster analysis, shorter experimental cycles, and reduced time to discovery.[21]

- *Viacom*, a global media company, found that the traditional data warehousing approach was too slow for its business. Traditional data warehousing calls for developers to structure and model data before it is released to end users. Instead, Viacom now uses Spark and Databricks running on Amazon Web Services to deliver "just-in-time" data warehouses. In this approach, end users define structure in the context of problems they needs to solve; data quality issues are resolved as they surface; developers and end users collaborate to answer business questions interactively. The end result: faster time to value and increased engagement of business users with the data.[22]

- *The Weather Company* (TWC) aggregates weather information from government agencies and distributes it to end users. Every day, it handles about 30 billion API requests from 120 million active mobile users, who generate 360 petabytes of traffic. TWC uses Spark Streaming, Cassandra, Parquet, and Spark SQL, all operating in the AWS cloud, to provide executives and managers with a real-time self-service platform for business intelligence and data science.[23]

Apache Arrow

In Chapter Four, we discussed the Apache Drill project, a schema-free SQL query engine built on the Google Dremel framework. The Apache Software Foundation accepted Drill as an Incubator project in September 2012.

[21] http://www.slideshare.net/SparkSummit/escaping-flatland-interactive-highdimensional-data-analysis-in-drug-discovery-using-spark-by-josh-snyder-victor-hong-and-laurent-galafassi
[22] http://www.slideshare.net/SparkSummit/building-a-just-in-time-data-warehouse-by-dan-morris-and-jason-pohl
[23] http://www.slideshare.net/SparkSummit/lambda-at-weather-scale-by-robbie-strickland

While working toward a first release in September 2013, Drill's developers identified a need to represent complex columnar data in memory. There were existing methods to represent structured data with a predefined schema in memory, and the Apache Parquet project had developed a way to represent complex columnar data on disk, but neither format met the needs of Drill's developers. To address this gap, the Drill team developed a data structure called Value Vectors.

Recognizing that Value Vectors meet the needs of other data processing engines, in February 2016, the Apache Software Foundation announced Apache Arrow as a top-level project, bypassing the standard Incubator process. Committers to the project include developers from other Apache projects such as Calcite, Cassandra, Drill, Hadoop, HBase, Ibis, Impala, Kudu, Pandas, Parquet, Phoenix, Spark, and Storm.

Apache Arrow enables execution engines like Spark to take advantage of the latest operations included in modern processors, for fast analytical data processing. Columnar layout of data also allows for a better use of CPU caches by placing all data relevant to a column operation in as compact of a format as possible.

A standard format allows applications to share data seamlessly. At present, each execution engine represents data in memory in its own way, so any handoff from one engine to another requires a time-consuming conversion. Standardizing the way that engines represent data in memory simplifies integration and speeds processing.

Apache Arrow software is available under the Apache License v2.0.

Dremio, a startup led by Jacques Nadeau, chair of the Apache Drill and Apache Arrow Project Management Committees, leads development. In September 2015, Dremio announced[24] an initial funding round of $10 million.

Alluxio

Alluxio (previously named Tachyon) is a distributed storage system that supports fault-tolerant data sharing at the speed of memory across cluster jobs. The Alluxio software resides between computation frameworks (including Spark, MapReduce, Flink, Zeppelin, HBase, and Presto) and storage systems (such as HDFS, Amazon S3, GlusterFS, OpenStack Swift, and NFS). Existing MapReduce and Spark programs run much faster on top of Alluxio without code changes.

---

[24]http://venturebeat.com/2015/09/25/apache-drill-gurus-at-dremio-raise-more-than-10m-from-redpoint-and-lightspeed/

A series of papers published by Haoyuan Li, Ali Ghodsi, Matei Zaharia, Scott Shenker, and Ion Stoica of University of California's AMPLab define the theoretical framework for Alluxio.[25] Haoyuan Li built the first version during Christmas of 2012 and released the code to open source in April 2013. Since then, Alluxio has attracted more than 200 contributors from over 50 companies and almost 200,000 lines of code, mostly written in Java.

Alluxio reports production deployments with hundreds of machines. Examples of Alluxio in action include:

- *Baidu*, the largest Chinese language search engine, runs Spark SQL queries 30 times faster with Alluxio.[26]

- *Barclays* found that Alluxio reduced runtime for a complex Spark workflow from hours to seconds.[27]

In March, 2015, Li formed Alluxio Inc. (then named Tachyon Nexus) to commercialize the software and announced an initial funding round of $7.5 million from Andreessen Horowitz.[28]

Apache Ignite

Apache Ignite is an open source project based on a code base donated to the Apache Software Foundation by GridGain Systems in October 2014. GridGain remains engaged in the project, with company executives holding the PMC Chair and multiple seats on the PMC. Operating on an open core business model, GridGain continues to offer commercially licensed value-added versions of the software.

Ignite combines a fault-tolerant ACID-compliant in-memory key-value store with tools for managing data. An embedded SQL engine supports ANSI-99 syntax. The project also includes:

- A fault-tolerant framework for implicitly parallel program execution.

- Scalable and fault-tolerant processing for continuous never-ending streams of data.

- High-performance cluster-wide messaging to exchange data among nodes.

---

[25]http://www.cs.berkeley.edu/~haoyuan/papers/2013_ladis_tachyon.pdf
http://www.cs.berkeley.edu/~haoyuan/papers/2014_EECS_tachyon.pdf
http://www.cs.berkeley.edu/~haoyuan/papers/2014_socc_tachyon.pdf
[26]http://www.alluxio.com/assets/uploads/2016/02/Baidu-Case-Study.pdf
[27]https://dzone.com/articles/Accelerate-In-Memory-Processing-with-Spark-from-Hours-to-Seconds-With-Tachyon
[28]http://blogs.wsj.com/venturecapital/2015/03/17/andreessen-horowitz-invests-7-5m-in-big-data-startup-tachyon/

Ignite can serve as a database cache, enabling users to keep the most frequently accessed data in memory. For this purpose, it offers write-through, read-through, and write-behind capability.

The Ignite File System (IGFS) is a distributed in-memory file system that works like HDFS, but in memory. IGFS splits data from each file into separate data blocks and stores them in a distributed in-memory cache, using a hashing function to determine file location. IGFS can be deployed by itself, or on top of HDFS, in which case it becomes a caching layer for HDFS (similar to Alluxio), with write-through and read-through capabilities. Apache Ignite includes an in-memory implementation of MapReduce.

Ignite provides an implementation of Spark RDDs which enable Spark jobs to share objects in memory, within the same Spark application or between different Spark applications. Running Spark SQL queries using the Ignite RDDs is faster than running them directly with the data, primarily because Ignite indexes the data in memory.

Apache Ignite runs standalone, in a cluster, within Docker containers, and Apache Mesos and under YARN. It has native integration with Amazon Web Services cloud and the Google Compute Engine.

GridGain Systems offers Professional and Enterprise Editions of Apache Ignite, for which it provides technical support. The Professional Edition is a binary build of the software, and it includes bug fixes not yet released in the open source version. The Enterprise Edition includes a capability to port in-memory objects across platforms, a tool to manage and monitor the environment, network segmentation, a recoverable local store, rolling production updates, and data center replication.

The New In-Memory Analytics

Consistent with the theory of disruptive innovation, the industry leaders in data warehousing have largely assimilated in-memory technology and incorporated it into their own offerings. As the cost of memory has declined absolutely, in-memory databases in various forms have become mainstream, embedded in the offerings of Oracle, SAP, IBM, Microsoft, and Teradata.

Startups featuring in-memory databases licensed under a commercial model, including EXASOL and Kognitio, have not disrupted the market. (Both vendors have been in business for at least 15 years.) On the other hand, NewSQL vendors MemSQL and VoltDB, both of which operate under an open core business model, have succeeded in tapping new Hybrid Transactional/ Analytical Processing (HTAP) use cases and demonstrate commensurate growth.

The greatest potential for market disruption comes from Apache Spark and related open source projects, which bring the power and speed of in-memory analytics to Hadoop and NoSQL. This disruptive power is not lost on the open source community, which has responded by making Spark the most active project in Big Data today.

Streaming Analytics

Insight from Data in Motion

Streaming analytics is the application of analytic operations to streaming data for applications such as:

- Algorithmic trading
- Customer interaction management
- Intelligence and surveillance
- Patient monitoring
- Supply chain optimization
- Network monitoring
- Oil and gas optimization
- Vehicle tracking and route monitoring

Market research firm *Markets and Markets* estimates[1] total spending on streaming analytics of $502 million in 2015, and predicts a 31% growth rate through 2020. While this is a sizeable market in its own right, keep in mind that it is only about 1% of worldwide spending on business analytics software.[2]

---

[1]http://www.marketsandmarkets.com/PressReleases/streaming-analytics.asp
[2]https://www.idc.com/getdoc.jsp?containerId=257402

© Thomas W. Dinsmore 2016
T.W. Dinsmore, *Disruptive Analytics*, DOI 10.1007/978-1-4842-1311-7_6

In this book, we use the term "real-time analytics" to describe a specific type of operation, described below. We use the term "streaming analytics" to describe the technology used in real-time analytics, and "low-latency analytics" to describe a desired outcome, minimized latency. The distinctions are important; while most examples of real-time analytics are low-latency, the reverse is not necessarily true.

Gartner defines "real-time" computing as:

> *The description for an operating system that responds to an external event within a short and predictable time frame. Unlike a batch or time-sharing operating system, a real-time operating system provides services or control to independent ongoing physical processes. It typically has interrupt capabilities (so that a less important task can be put aside) and a priority-scheduling management scheme.*[3]

In other words, real-time analytics:

1. Must be completed in a defined window of time.

2. Are initiated by an external process, such as incoming data.

3. Have priority over other operations.

A real-time operation is *not* "zero-latency". Since any operation takes some time to complete, all operations entail some latency, even if only a millisecond. However, since real-time operations are initiated by an external process (the second criterion cited here), and have priority over other operations, they experience no waiting time *attributable to queueing*.

Real-time analytics aren't distinguished by any absolute level of latency or time window for execution; the use case defines the time window. In an electronic market, an ultra-low latency trade is completed in less than one millisecond.[4] In the field of plate tectonics an annual measure could qualify as "real-time".

Ignoring this distinction produces odd debates over the precise definition for "real-time":

- Vivek Ranadivé, founder of TIBCO, a company whose DNA is real-time analytics, writes about "the two-second advantage".[5]

- Forrester analyst Mike Gualtieri asserts that "real-time" is anything less than one second latency.[6]

---

[3] http://www.gartner.com/it-glossary/real-time
[4] http://home.business.utah.edu/finmh/moallemi.pdf
[5] http://www.amazon.com/Two-Second-Advantage-Succeed-Anticipating-Future-
-Just/dp/0307887650/ref=sr_1_1?s=books&ie=UTF8&qid=1462718451&sr=1-1&key
words=the+two+second+advantage
[6] https://spark-summit.org/east-2016/events/5-reasons-enterprise-
adoption-of-spark-is-unstoppable/

- Capitol One's Slim Baltagi argues[7] that Apache Spark isn't suitable for real-time analytics because it can "only" reduce latency to half a second.

Does real-time mean "two seconds or less," "less than one second," or "less than half a second"? The answer is possibly all of the above, and possibly none of the above. It all depends on the use case.

The potential of real-time analytics inspires hyperbole. Eric Woods of Navigant Research writes:

> How well we deliver on the goal of real-time analytics will tell us much about the real level of maturity of our systems and our managerial structures.[8]

He wrote that in 2002.

Advocates for real-time analytics overstate its importance in the business analytics universe. Admiral Husband E. Kimmel, Commander in Chief of the United States Pacific Fleet at Pearl Harbor on December 7, 1941, had no shortage of real-time information about the Japanese attack. What he needed was a prediction.

We begin the chapter with a short history of streaming analytics, followed by a review of streaming fundamentals. In the third section of the chapter, we review popular streaming data sources, such as Apache Kafka and Amazon Kinesis, followed by a survey of the top open source streaming engines. We close the chapter with some examples of streaming analytics in action, and some observations about the economics of streaming.

A Short History of Streaming Analytics

As discussed in Chapter Two, the digital transformation of business creates new opportunities for analytics. In the 1980s, financial markets were just beginning to transition from the open outcry system, an auction system based on human traders in one location, to electronic trading. NASDAQ was the first U.S. market to do so, followed by the Chicago Mercantile Exchange.

Vivek Ranadivé, a 28-year old graduate of MIT and the Harvard Business School, saw an opportunity to speed the integration of systems, and founded Teknekron Software Systems in 1986. Teknekron developed a software bus to

---

[7]http://www.slideshare.net/sbaltagi/flink-vs-spark
[8]http://www.computerweekly.com/opinion/Why-real-time-CRM-analytics-is-hot

link software programs and transfer information between them at high speed, marketing the product under the Information Bus trademark. Teknekron described[9] the product as:

> *Computer software to aid the process whereby one computer obtains data from another computer by receiving requests for data by subject matter, mapping the subject to a particular computer on a network which can supply that data and performing any necessary format conversion operations between incompatible formats so that the data received on the subject can be used by the requesting computer.*

Goldman Sachs engaged[10] Teknekron to develop a stock trading system. Fidelity Investments, First Interstate Bank, and Salomon Brothers invested in systems based on Teknekron's Information Bus to integrate and deliver stock quotes, news, and other financial information to traders. Since most of the financial markets in the 1980s continued to operate through open outcry, investors and brokers focused on consolidating information for human traders rather than algorithmic trading.

Meanwhile, in the credit card business, fraud was a growing problem. Worldwide deployment of the Visa and MasterCard interchange systems created a growing need for real-time fraud detection. In 1992, HNC Software introduced its Falcon application, as discussed in Chapter Two. Falcon analyzed individual credit card transactions as they were presented for authorization, flagging suspects within the time window dictated by Visa/MasterCard rules.

HNC built Falcon on mainstream client-server technology (with an indexed file system for fast profile lookups). Falcon could analyze and disposition authorization requests at the rate of 120 transactions per second, which was more than adequate at the time.[11] Within ten years, Falcon handled 80% of the global Visa/MasterCard transaction authorizations.

Growing use of the World Wide Web created opportunities for real-time customer interaction management. While working at the University of Paris in the early 1990s, Dr. Khai Minh Pham developed a predictive modeling technique he called Agent Network Technology[12] (ANT). In 1994, he founded DataMind, offering a product branded as DataCruncher; the company launched in 1996 with $4.7 million in venture capital and a patent on ANT.

---

[9]http://www.trademarkia.com/information-bus-74089524.html
[10]http://www.risk.net/operational-risk-and-regulation/feature/1507883/goldman-to-roll-out-teknekron-middleware-transaction-platform
[11]http://www.amazon.com/Business-Applications-Neural-Networks-State/dp/9810240899
[12]http://www.cbronline.com/news/datamind_boosts_business_intelligence

In 1998, DataMind rebranded itself as Rightpoint and repositioned its product as a platform for real-time customer interaction management. Leveraging the "self-learning" ANT technology, Rightpoint offered customers capabilities for clickstream tracking, collaborative filtering, and real-time profiling to drive offer recommendations and personalized content. CRM vendor E.piphany Inc. acquired[13] Rightpoint for $393 million ($562 million in 2016 dollars) in 1999.

While Khai Minh Pham developed ANT and launched DataMind, Stanford's David Luckham proposed a research project in discrete event simulation to the Advanced Research Projects Agency.[14] A key element of this project (named Rapide) was the ability to model patterns of concurrent events—in other words, the ability to infer higher-level events from many discrete events closely spaced in time.

Beginning in 1993, Luckham and his team developed the Rapide language and techniques to infer patterns or high-level events from streams of messages about more atomic events. This work culminated in 1998 with the publication of *Complex Event Processing in Distributed Systems*.[15]

Simultaneously, at Cambridge University, John Bates[16] developed a theory of complex event inference that was similar to Luckham's. Lacking tools to describe and handle patterns of complex events, Bates and his team developed the necessary software. In 1999 Bates and Giles Nelson, a colleague, founded a company branded as APAMA to commercialize the software, targeting traders and investors in financial markets. (They sold[17] the company to Progress Software in 2005 for $25 million.)

In the same year, a team of former Cray executives founded Aleri Group. Aleri offered a high-performance platform built on a vector database[18], a technology that offered the throughput needed to manage and analyze high-speed high-volume trading data.

Vivek Ranadivé had sold[19] Teknekron to Reuters in 1993 for $125 million ($206 million in 2016 dollars); with the proceeds, he founded TIBCO Software in 1997. Unlike APAMA and Aleri, TIBCO focused on middleware, the infrastructure that ties systems together and makes low-latency communications possible. Timed perfectly for the late 1990s internet boom, the company grew rapidly; revenue increased from $53 million in 1998 to $96 million in 1999. TIBCO's initial public offering at $10 in 1999 valued the company at more than $300 million.

---

[13]http://www.nytimes.com/1999/11/17/business/company-news-epiphany-agrees-to-acquirerightpoint.html.
[14]Renamed the Defense Advanced Research Projects Agency in March 1996.
[15]http://citeseerx.ist.psu.edu/viewdoc/download?doi=10.1.1.56.876&rep=rep1&type=pdf
[16]http://www.softwareag.com/special/thingalytics/john-bates.html
[17]https://www.finextra.com/news/fullstory.aspx?newsitemid=13477
[18]http://www.computerweekly.com/feature/What-can-science-do-for-IT
[19]http://www.nytimes.com/1993/12/18/business/company-news-reuters-is-buying-teknekron.html

Wall Street fell in love with TIBCO's stock, driving it up to 12 times the IPO price by the summer of 2000. Then, in the wake of the internet bust, the stock fell to single digits by 2001.

In that year, Michael Stonebraker of MIT, together with scientists from Brandeis and Brown, started work on the Aurora project for data stream management. The team designed Aurora to handle large numbers of asynchronous push-based data streams (in contrast to relational databases, from which users "pull" data with discrete requests). Aurora represented streaming processes as a directed graph that mapped the flow of data from sources, through streaming operators, and then to consuming applications.[20]

Aurora users built continuously operating queries with standard filtering, mapping, windowing, and join operations. Windowed operations supported timeout and slack parameters enabling the engine to handle slow and out-of-order operations.

Stonebraker and others founded StreamBase Systems in 2003 to commercialize an enterprise-grade version of the Aurora engine. Like Aurora, StreamBase uniquely integrated streaming and SQL operations into a single platform. Backed with $5 million in venture capital, StreamBase released[21] its first products in August 2004 and closed[22] an $11 million "B" round in January 2005, intending to sell to investment banks, hedge funds, and government agencies.

Financial markets transformed rapidly in this period. The U.S. Securities and Exchange Commission authorized electronic trading in regulated securities in 1998. Electronic trading created opportunities for algorithmic trading to detect and execute trades exploiting short-lived opportunities. These included arbitrage between markets, arbitrage between indexes and the underlying stocks, and licit and illicit tactics.

Reducing latency in market trades makes markets more efficient. However, trading itself is a zero-sum game, with benefits accruing to the trader that can accumulate and act on information faster than all other traders. Between 1999 and 2010, market participants invested heavily in the tools and infrastructure needed to drive latency out of their trading operations, in a kind of an arms race. The time needed to complete trades declined to milliseconds and even microseconds. In 2010, an executive of the Bank of England predicted that trading time would decrease to nanoseconds.[23]

---

[20]http://cs.brown.edu/research/aurora/sigmoddemo.pdf
[21]http://www.siliconinvestor.com/readmsgs.aspx?subjectid=34520&msgnum=23107&batchsize=10&batchtype=Next
[22]http://www.prnewswire.com/news-releases/streambase-systems-secures-11-million-to-expand-sales-and-marketing-activities-66325312.html
[23]http://www.bis.org/review/r100909e.pdf

The key players in streaming analytics profited from growing interest in low-latency analytics. Through the 2000s, TIBCO steadily delivered low latency infrastructure to an expanding list of industries: mobile telecommunications; airlines, for baggage handling, ticketing, and check-in; insurance, for handling claims; and to the gaming industry. Amazon.com adopted TIBCO middleware to support its recommendation engine, and FedEx deployed it for package tracking. Through 2009, TIBCO's revenue grew by double digits.

In the same period, under management by Progress Software, APAMA increased its presence on Wall Street for algorithmic trading, as well as in retail banking, telecommunications, logistics, government, energy, and manufacturing. APAMA's strong point was a visual interface that made it easy for business analysts to set up and run streaming applications.

The streaming analytics industry began to consolidate in the late 2000s. In 2009, Aleri merged with Coral8, a competing CEP vendor. The combined entity offered a suite of software for liquidity management, low latency trading across markets, low latency risk management, and stress testing. Database vendor Sybase acquired the assets of Aleri in March 2010 and enterprise software vendor SAP acquired[24] Sybase two months later.

IBM entered the streaming analytics market in 2009. IBM Research spent years developing[25] the streaming platform it called "System S". Touting the system under the neologism "perpetual analytics"—a concept that did not stick— IBM released[26] the product branded as IBM Infosphere Streams.

For high performance and throughput, IBM designed Streams to distribute workload over clustered servers. IBM invented[27] two new languages for streaming analytics—Stream Processing Application Declarative Engine (SPADE[28]) and Mashup Automation with Runtime Invocation & Orchestration (MARIO)—neither of these has gained any acceptance outside of IBM. Infosphere Streams scored well on performance and scalability tests, and it supported comprehensive operators and development tools in an Eclipse-based integrated development environment.

---

[24]http://blogs.forrester.com/holger_kisker/10-05-13-sap_acquires_sybase_%E2%80%93_what%E2%80%99s_strategic_intent_behind_deal
[25]http://www.geek.com/chips/ibm-releases-system-s-real-time-stream-computing-analysis-and-reporting-773531/
[26]http://www-03.ibm.com/press/us/en/pressrelease/27508.wss
[27]http://www.enterrasolutions.com/media/docs/2012/01/SystemS_2008-1001.pdf
[28]http://cs.ucsb.edu/~ckrintz/papers/gedik_et_al_2008.pdf

Despite its technical strengths, StreamBase struggled to compete in the narrow niche of real-time analytics. After stating that its 2005 venture funding would be its last private round, StreamBase closed another round in 2007, then accepted a "down" round in 2009 shortly after the departure of CEO Barry S. Morris. In 2013, TIBCO acquired StreamBase Systems for $49.7 million.[29]

Two days later, Progress Software sold APAMA to Software AG for $44 million.[30] Software AG supplemented the core APAMA software with a suite of tools for order routing, pre-trade risk, and other building blocks for capital markets solutions; a separate scoring engine that works with predictive models trained offline and imported through PMML; and a dashboarding application.

As of 2016, Software AG, IBM, SAP, TIBCO, and Oracle dominate the commercial market for streaming analytics. (Oracle entered the market through its acquisition of BEA Systems in 2008.) Forrester rates[31] all five as "leaders" in its annual market survey, together with startups SQLstream and DataTorrent.

Fundamentals of Streaming Analytics

In this section, we cover the basics of streaming analytics: the definition of stream processing, streaming operations, Complex Event Processing, streaming machine learning, and anomaly detection.

Stream Processing

The term "streaming analytics" combines two concepts: *streaming data processing* and *low-latency analytics*. Streaming data processing is a systems paradigm or architecture that processes data as it arrives. Low-latency analytics is a desired outcome, the delivery of insight about events as soon as possible after those events occur in the real world. Organizations use streaming data processing as a means to accomplish the end of low-latency analytics.

We contrast streaming data processing with *batch processing*. Under batch processing, programs work with finite sets of data gathered in discrete sets, or batches. The program runs until it finishes handling the data included in the batch, then terminates.

---

[29]http://www.sec.gov/Archives/edgar/data/1085280/000108528014000020/tibx1130201310k.htm
[30]https://www.sec.gov/Archives/edgar/data/876167/000087616714000013/a201310-kmaster.htm
[31]https://www.forrester.com/report/The+Forrester+Wave+Big+Data+Streaming+Analytics+Q1+2016/-/E-RES129023

A batch process can run on a fixed schedule, such as every night at midnight. It can also run whenever the accumulated data reaches a certain threshold, or it can run on an ad hoc schedule under the manual control of an operator.

Batch processing is efficient, but it builds latency into the process. In a batch process, the total latency equals the amount of time an arriving record waits in a queue, plus the time needed to perform operations on the data. This latency can be highly visible when the information or insight is critical for the operations of the organization.

Streaming data processing works in a different manner. In streaming data processing, programs handle data as it arrives, one unit at a time. Once the data stream starts, processing continues as fast as the data arrives and continues without a predetermined end.

Since a streaming process runs continuously, without startup or cleanup, it must be fault-tolerant, with the ability to transfer processing from a failed node to a working node, and the ability to reconstruct data when a process fails.

In addition to reduced latency, a well-designed streaming process spreads workload over the day. In contrast, a batch process requires workload "spikes" that may be more difficult to manage. Organizations can schedule batch operations to run in off-peak periods; but as organizations use cloud computing and virtualization to shift workloads, there are fewer off-peak periods when infrastructure is idle.

Streaming systems must enforce one of three processing semantics:

- *At-least-once* processing ensures that no messages sent by the source system will be omitted by the receiving system.

- *At-most-once* processing ensures that no messages will be duplicated in the receiving system.

- *Exactly-once* processing ensures that each message sent by the source system is captured in the receiving system once and only once.

We use the term "streaming analytics engine" to characterize streaming platforms with the ability to perform analytic operations (defined later in this chapter) and *without* a storage capability. Thus, a streaming analytics engine accepts streaming input, performs an operation, and passes the result to some other application for storage or use.

The ability to perform analytic operations distinguishes a streaming analytics engine from a streaming data source, which simply collects and forwards streaming data. However, we expect the two categories to converge, as popular streaming data sources add analytics capabilities.

Databases can also have streaming capabilities. However, in a 2004 paper[32] MIT's Michael Stonebraker detailed the important differences between a relational database and a purpose-built streaming engine. Stonebraker's team compared throughput for a workflow with 22 operators on StreamBase and on a leading relational database. StreamBase processed data at the rate of 160,000 records per second; by comparison, the team could achieve only 900 records per second with the relational database.

Stonebraker attributed the extreme performance difference to the relational database's mandatory storage operation for incoming records. StreamBase made the initial storage optional, so that incoming records could be processed in memory, then either stored or passed to another application. In short, the *sequence* of operations matters; a database that stores data first, then performs calculations will appear to be slower than an in-memory engine that simply performs calculations and passes data to another application for storage. Of course, the total processing time from data receipt to storage matters; but for some applications, it makes sense to perform calculations first and store the data in background.

Databases have evolved a great deal since 2004; many today permit in-memory pre-processing before storage. Moreover, the growth of in-memory databases, covered in Chapter Five, tends to blur the distinction between pure streaming engines and databases. Startups MemSQL and VoltDB, for example, position their in-memory databases in the market for streaming analytics.

Streaming Operations

A conventional relational database processes a query until it reaches end-of-table indicators for all tables referenced in the query. In a streaming database, there is no equivalent end-of-table concept, since the database updates continually as new records arrive.

Many of the operations analysts seek to perform on streams are the same as those they perform on static tables. Some, however, are unique to streaming data: they include joining streams, aggregations, filtering, windows, and alerts.

Joining Streams. For insight, analysts may need to join multiple streams to one another. For example, a vehicle fleet operator may have streams of data arriving for each vehicle in the fleet; to monitor fleet-level statistics, all of the individual streams must be joined into a single stream.

---

[32]https://cs.brown.edu/~ugur/fits_all.pdf

For context, analysts may also need to join streams to static tables. For example, suppose that we are working with a stream of transactions posted by hundreds of retail stores, which we want to group by region. The transaction records in the incoming stream have a store code but not a region code; for that, we must join the stream to a static store master table to capture the region code.

Aggregations. A key capability of streaming SQL is to the ability to compute and retain aggregates on the incoming stream. For example, we may want to compute a cumulative count and sum of transactions as they arrive. Aggregations are useful when combined with windowing, so the computed measures correspond to statistics for discrete and finite time intervals.

Filtering. Filtering a stream of data is conceptually similar to filtering a static data set. We filter for two reasons:

- To remove noise and irrelevant data.

- To limit the scope of the analysis to a specific subset of the data.

In the first case, the stream of data may include test records, incomplete transactions, miscoded data, or other kinds of "garbage" that simply adds noise to our analysis.

Business questions rarely require the entire universe of data available. Instead, we typically seek information about specific products, stores, people, customers, geographies, and time periods (or complex combinations of all of these attributes).

Windowing. Streaming data arrives continuously at arbitrary time intervals. For insight and analysis, however, end users want to see statistics for fixed time intervals: seconds, minutes, hours, or some other interval. Windowing functions enable users to define a time period, or window, and the data to include in the window. Analysts can use statistics aggregated through windowing for cumulative totals, moving averages, and other more sophisticated analysis.

Alerts. Alerts are arguably the most important streaming operator. Streaming analytics theory holds that (a) information becomes less valuable as it ages, and (b) information is valuable only if it is actionable. Thus, defining and tuning alerts is a central task for any streaming analytics system.

There are three kinds of rule-based alerts, in increasing complexity:

- Alerts based on fixed rules, universally applied. For example, "select all transactions with an amount greater than $1,000".

- Alerts based on rules that are differentiated by groups or entities within the population. For example, "select all transactions greater than two times the mean transaction value for this customer".

- Alerts based on rules that are differentiated by groups or entities over time. For example, "select all transactions greater than two times the mean transaction value for this customer in the past month".

In any streaming system, defining and tuning alerts require careful balancing of false positives and negatives. Alerts drive actions, such as fraud investigations, transaction declines, or tax audits; some actions are expensive, or they can adversely impact customer goodwill.

False positives are like false alarms: the system focuses attention on a transaction that, upon further investigation, requires no action. False negatives are lost opportunities; the system fails to focus attention on a transaction that should have been acted upon.

If we define alerts too narrowly, the system produces too few alerts, and there are many false negatives. In a credit card fraud detection system, for example, the result will be high fraud losses.

On the other hand, if we define alerts too broadly, the system produces too many alerts, many of which are false alarms; the system loses credibility with its users. In a credit card fraud detection system, the result will be angry customers and overworked investigators.

In practice, alerts based on streaming data must be carefully designed, developed, tested, validated, tuned, and monitored. These needs drive many of the supporting tools and capabilities of commercial streaming analytics platforms.

Events in a stream rarely provide useful insight in isolation; we need context to distinguish important events. Consider the following example:

- At 11:41 am on Saturday, John Doe presents his credit card at a particular store in Chicago for a $500 purchase.

If we are interested in detecting credit card fraud, this transaction alone tells us very little. But now consider the following context:

- In the past 12 months, a high percentage of the credit card transactions at this store were fraudulent.

- For stores of this type, transactions of more than $100 have a higher incidence of fraud.

Adding some historical information about the merchant and merchant category raises our concern about the transaction. Checking Mr. Doe's profile, we discover that he lives in Philadelphia, which might be enough to trigger a request to the store cashier to verify the customer's identity. Now consider the following additional fact:

- At 11:01 am today, Mr. Doe presented the same card at a gas station in Philadelphia.

Since it is not possible for a customer to present the same credit card in two widely separated cities, we now know that at least one of the cards is fraudulent, and we can take further action.

The example demonstrates a key principle of streaming analytics: for insight, we must combine a streaming fact with other streaming facts and with stored contextual data. Operations on individual streaming facts alone produce trivial results.

In a related example, suppose that we are interested in customer loyalty and retention for an online bank. Every day for the past three years, Jane Doe has logged in to check her balance. On March 1, she does not log in. Like Sherlock Holmes' dog that did not bark[33], it is the *absence* of a transaction that matters, and we are only able to understand this because we combine streaming facts with history.

We cite these two examples to demonstrate that for most business analytics, streaming data cannot be separated from historical data; both must be used together. Reflecting this key point, tools for the analysis of streaming data and static data are converging; instead of distinct tooling for streaming data, we see streaming *operators* within analysis tools that can work with both types of data.

Complex Event Processing

We noted earlier in this chapter, in the historical survey, that Complex Event Processing (CEP) emerged in the 1990s. The streaming analytics startups that emerged at this time generally featured CEP as an organizing principle for interactions with users.

---

[33]http://www.thenational.ae/business/the-curious-case-of-the-dog-that-did-not-bark

Gartner defines CEP as:

> *A kind of computing in which incoming data about events is distilled into more useful, higher level "complex" event data that provides insight into what is happening. CEP is event-driven because the computation is triggered by the receipt of event data. CEP is used for highly demanding, continuous-intelligence applications that enhance situation awareness and support real-time analytics.*[34]

Interest in CEP peaked around 2010, as shown in Figure 6-1.

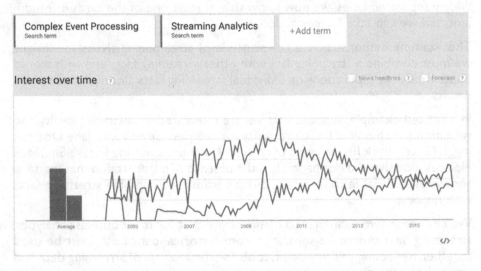

Figure 6-1. Google search interest in CEP (Source: Google Trends)

CEP is an analytic framework that enables inference of high-level patterns or events from multiple streaming data sources. For example, in capital markets a trader might seek to use CEP to infer buy and sell signals for a security from streaming news feeds, text messages, social media, market feeds, weather reports, and other data, collectively called an *event cloud*.

Suppose, for example, that we have a large number of sensors mounted on a Formula One racing car that measure such things as oil pressure, water pressure, exhaust particulates, and power output every fraction of a second. We want to detect a blown engine as early as possible, so we can automatically shut down other systems to avoid damage and inform the driver to steer the car to safety.

From analysis, we know that a rapid drop in oil pressure and water pressure combined with reduced engine power output and increased particulates in the exhaust mean that the engine is blown. With CEP, we can model those relationships; then, we can use CEP software to build an automatic shutdown procedure using data streaming from the sensors.

---

[34]http://www.gartner.com/it-glossary/complex-event-processing

CEP is not an algorithm, but a conceptual model for a class of problems where useful patterns require combining information across different sources of streaming data, and where events of interest are defined in time. CEP does not tell the user what the relationships *are* among events; it simply allows the user to describe them. In this respect, CEP is comparable to SQL; it enables the user to express a pattern and generate data accordingly, but does not help the user discover patterns.

In theory, analysts can use machine learning to infer relationships between high-level events and detailed data. In practice, most existing applications depend more on rule-based inference, because rules are easier for business users to understand.

Streaming Machine Learning

More often than not, when managers speak about machine learning with streaming data, they mean *scoring* with streaming data. In other words, they want to use a predictive model trained with static data to generate predictions with streaming data. In low latency scoring, the model itself remains stable; we simply seek to apply a stable model to produce a prediction with minimal latency. Organizations use such predictions in a variety of automated decisions, such as scoring live credit card transactions for fraud risk.

Most streaming engines and low latency decision engines can ingest PMML models, and can also support predictive model pipelines as custom code. From the model developer's perspective, training predictive models that will be deployed with streaming data is no different from any other deployment scenario, although the model developer may need to be mindful of what data will actually be available in a low latency environment.

Evolutionary machine learning algorithms that learn continuously from streaming data represent an entirely different type of model. An evolutionary model is appropriate when the process we seek to model is not stable over time. Such algorithms are rarely, if ever, used to support business decisions, due to legal, regulatory, and management concerns. Actual applications at present tend to be limited to experimental use cases, or where the analytics will be used for insight and discovery.

The two key issues for evolutionary models are data recency and the time window. By recency, we mean how quickly new transactions enter the model training data and the model itself adjusts to the new observation. The time window of the model is the amount of history included in the training data; models can work with a sliding window (such as the last 24 months of data), a fixed window (all data captured after January 1, 2014), or all data ever captured.

Evolutionary models with very long time windows produce results that are very similar to static models. Static models that are updated frequently produce results that are very similar to evolutionary models unless the process we seek to model is highly unstable—in which case the value of predictive modeling itself is called into question.

On paper, at least, there are incremental versions of support vector machines[35], neural networks,[36] and Bayesian networks[37]. There are actual implementations of incremental versions of k-means clustering[38] and linear regression[39] in Apache Spark.

Anomaly Detection

Anomaly detection is the identification of items in a stream that do not conform to an expected pattern or to other items in the stream. Technically, anomaly detection is not limited to stream processing and can be applied to batches of data as well. In practice, however, organizations use anomaly detection for low-latency applications, such as network security, where the goal is to evaluate events as soon as possible.

Like CEP, anomaly detection is neither a precisely defined technique nor a specific use case; it is a generic application that can be applied to a broad range of business problems, including fraud detection, health care quality assurance, operations management, and security applications. Machine learning techniques suitable for anomaly detection include:

- Supervised learning techniques, where anomalous events are well-defined and we have a set of examples we can use to train a model.

- Unsupervised learning techniques, where we cannot define anomalies in advance and simply seek to identify cases that are different.

Practitioners have successfully used k-nearest neighbor, support vector machines, neural networks, clustering, association rules, and ensemble models to build anomaly detection applications. Anomaly detection systems based on unsupervised learning generate alerts and route them to human analysts for investigation and disposition.

---

[35]http://www.isn.ucsd.edu/pubs/nips00_inc.pdf
[36]ftp://ftp.sas.com/pub/neural/FAQ2.html#A_styles_batch_vs_inc
[37]http://research.microsoft.com/apps/pubs/default.aspx?id=69588
[38]http://spark.apache.org/docs/latest/mllib-clustering.html#streaming-k-means
[39]http://spark.apache.org/docs/latest/mllib-linear-methods.html#streaming-linear-regression

Streaming Data Sources

The streaming data platforms detailed in this section differ from the low-latency analytics platforms covered later in this chapter because they lack analytics operators. Instead, they are designed to serve as brokers between source systems that generate data and analytic systems that consume data.

We note that the distinction between streaming data sources and streaming engines may blur in the future, as popular data sources add analytics capability. Amazon Web Services has announced Kinesis Analytics (planned availability is late 2016).

Apache ActiveMQ

Apache ActiveMQ is a Java-based open source message broker. ActiveMQ entered Apache incubation in December 2005 and graduated to top-level status in February 2007. ActiveMQ is widely used and embedded in at least 16 other Apache projects.

Red Hat distributes JBoss A-MQ, a commercially supported version of ActiveMQ. Several other companies offer training, consulting, and support for the generic open source version.

ActiveMQ's code base expanded[40] rapidly until late 2010 and has expanded slowly since then.

Apache Kafka

Apache Kafka is an open source message-brokering software project. A team at LinkedIn developed the original code; LinkedIn contributed the code to the Apache Software Foundation in 2011. Kafka graduated to top-level Apache status in October 2012.

Kafka offers very high throughput for streaming data. Individual Kafka servers ("brokers") can handle hundreds of megabytes of reads and writes per second from thousands of sources.

Kafka has a distributed scale-out architecture for durability and fault-tolerance. The software saves and replicates messages within the cluster, so it can reconstruct messages if a node fails.

---

[40]https://www.openhub.net/p/activemq

A single Kafka cluster can serve as the central data backbone for a large organization. The cluster can be expanded and contracted without downtime. Kafka partitions data streams and spreads them over a cluster of machines to support data streams that are too large for any single machine to handle.

Some of the ways that organizations use[41] Kafka include:

- LinkedIn uses Kafka to handle activity stream data and operational metrics and to power products such as LinkedIn Newsfeed and LinkedIn Today.

- Netflix uses[42] Kafka for its low-latency event processing pipeline.

- Spotify uses Kafka to ingest 20 terabytes of data daily as part of a log delivery system.

- Square uses Kafka to move systems events, including metrics and logs, through a low-latency pipeline to consuming systems like Splunk and Graphite.

- Cisco's OpenSOC[43] project seeks to develop an extensible and scalable advanced security analytics tool. OpenSOC uses Kafka to collect streaming data from traffic replicators and telemetry sources and pass the data along to Apache Storm for low-latency analytics.

- Retention Science uses Kafka to collect and handle clickstream data.

Kafka has a relatively small code base[44], but contributions have accelerated markedly since mid-2015.

Confluent, a startup founded in 2014, leads Kafka development and offers training, consulting, and commercial support.

Amazon Kinesis

Amazon Kinesis is an Amazon Web Services (AWS) platform for streaming data. The service enables users to load and analyze streaming data and to build streaming applications. The platform includes three services:

- Amazon Kinesis Firehose, a basic service for handling streaming data.

---

[41]https://cwiki.apache.org/confluence/display/KAFKA/Powered+By
[42]http://cdn.oreillystatic.com/en/assets/1/event/118/The%20Evolution%20
of%20Hadoop%20at%20Spotify-%20Through%20Failures%20and%20Pain%20
Presentation.pdf
[43]http://opensoc.github.io/
[44]https://www.openhub.net/p/apache-kafka

- Amazon Kinesis Analytics, a SQL service for streaming data (planned availability is late 2016).

- Amazon Kinesis Streams, a service for building applications that handle streaming data.

Amazon Kinesis Firehose captures and automatically loads streaming data into Amazon S3 and Amazon Redshift. AWS offers Firehouse as a managed service that scales up and down automatically to handle variation in data volumes with consistent throughput.

Amazon Kinesis Streams enables users to build custom applications that continuously capture and process data from sources such as web site click-streams, financial transactions, social media feeds, IT logs, and location-tracking events. AWS offers a library of prebuilt applications for common tasks such as building low-latency dashboards, alerts, dynamic pricing, and so forth. Users can also transmit from Kinesis to other AWS services such as S3, Redshift, Amazon Elastic Map Reduce (EMR), and AWS Lambda.

RabbitMQ

RabbitMQ is an open source message broker that implements a standard called the Advanced Message Queing Protocol (AMQP) on the Open Telecom Platform (OTP) clustering framework. Rabbit Technologies, Limited, a subsidiary of Pivotal Software, leads development and provides commercial support.

RabbitMQ has an active list of contributors[45] and a gradually expanding code base.

Streaming Analytics Platforms

As of early 2016, there are five Apache open source projects that support streaming analytics: Apache Apex, Apache Flink, Apache Samza, Apache Spark (Streaming), and Apache Storm. Apex, Flink, Samza, and Storm are "pure" streaming engines, while Spark is a general-purpose analytics platform with streaming capabilities. The number of projects reflects the rapidly growing interest in streaming among prospective users and contributors.

---

[45]https://www.openhub.net/p/rabbitmq

There are also streaming tools available in R, Python, and other open source libraries. We focus on the Apache projects because they all support fault-tolerant distributed computing.

Apache Apex

Apache Apex is an the open source version of a streaming and batch engine originally developed by DataTorrent, a commercial venture founded in 2015. Most of the code commits to Apex come from DataTorrent employees. Apex entered Apache incubator status in August 2015 and graduated[46] to top-level status in April 2016.

To supplement core Apex, its developers created the Malhar library, which includes operators that implement common business logic functions needed by customers who want to quickly develop applications. These include:

- Access to file systems, including HDFS, S3, and NFS.

- Integration with message brokers, including Kafka, ActiveMQ, RabbitMQ, JMS, and other systems.

- Database access, including connectors to MySQL, Cassandra, MongoDB, Redis, HBase, CouchDB, and other databases along with JDBC connectors

The Malhar library also includes common business logic patterns that help users reduce the time it takes to go into production.

In its proposal to the Apache Software Foundation, Apex, sponsors differentiate the project by arguing that applications written for non-Hadoop platforms typically require major rewrites to get them to work with Hadoop.

> *This rewriting creates a significant bottleneck in terms of resources (expertise), which in turn jeopardizes the viability of such an endeavor. It is hard enough to acquire Big Data expertise; demanding additional expertise to do a major code conversion makes it a very hard problem for projects to successfully migrate to Hadoop.*[47]

The Apex team reports no current production users. After growing rapidly until early 2014, the code base[48] has been largely static since then.

---

[46]http://apex.apache.org/announcements.html
[47]https://wiki.apache.org/incubator/ApexProposal
[48]https://www.openhub.net/p/apache_apex

Apache Flink

Apache Flink is an open source distributed dataflow engine written in Scala and Java. Flink's runtime supports batch and stream processing, as well as iterative algorithms.

Dataflow programming is an approach that models a program as a directed graph of the data flowing between operations. Dataflow programming focuses on data movement and models programs as a series of connections.

Flink does not have a storage system, so input data must be stored in a file system like HDFS or HBase, or it must come from a messaging system like Apache Kafka.

Researchers at the Technical University of Berlin, Humboldt University of Berlin, and the Hasso-Plattner Institute started a collaborative project called Stratosphere[49] in 2010. After three Stratosphere releases, the consortium donated a fork of the project to the Apache Software Foundation, which accepted the project for Incubator status in March 2014. In December 2014, Flink graduated to top-level status.

Some of the ways that organizations use Flink include:

- Capital One uses Flink for low-latency customer activity monitoring.

- Bouygues Telecom uses Flink for low-latency event processing and analytics.

- ResearchGate uses Flink for network analysis and duplicate detection.

Other applications include semantic Big Data analysis and inference for tax assessment and research on distributed graph analytics.

Flink includes several modular libraries, including:

- Gelly, a Graph API for Flink, with utilities that simplify the development of graph analysis applications.

- ML, a machine learning library.

- Table, an API that supports SQL-like expressions.

---

[49]http://stratosphere.eu/

DataArtisans, a startup based in Berlin, Germany, leads development, provides commercial support, and organizes the Flink Forward conference. Flink is not currently supported in any commercial Hadoop distribution.

Flink's code base[50] grew steadily until mid-2015, but has been largely flat since then.

Apache Samza

Apache Samza is a computational framework that offers fault-tolerant, durable, and scalable stateful stream processing with a simple API. It uses Kafka for messaging and runs under YARN. A team at LinkedIn developed Samza together with Kafka to support stream processing use cases. LinkedIn donated[51] the project to open source in 2013; it entered Apache incubator status in July 2013, and graduated to top level status in January 2015.

Some of the ways that organizations use Samza include:

- LinkedIn uses Samza to process tracking and service log data and for data ingestion pipelines.

- Intuit uses Samza to enrich events with more contextual data from various sources to aid operations personnel.

- Metamarkets uses Samza to transform and join low-latency event streams for interactive querying.

- Uber uses Samza to aggregate metrics, database updates, fraud detection, and root cause analysis.

- Netflix uses Samza to route more than 700 billion events per day from Kafka to Hive.

Other applications include low-latency analytics, multi-channel notification, security event log processing, low-latency monitoring of data streams from wearable sensors for healthcare management, and social media analysis.

Samza is not currently supported by any of the Hadoop vendors. Contributor activity[52] is low.

Apache Spark Streaming

We discussed Apache Spark's streaming capabilities in Chapter Five, under in-memory analytics.

---

[50]https://www.openhub.net/p/flink
[51]https://engineering.linkedin.com/data-streams/apache-samza-linkedins-low latency-stream-processing-framework
[52]https://www.openhub.net/p/samza

Apache Storm

Apache Storm is an open source low-latency computing system. A team at a startup named BackType wrote Storm in the Clojure programming language. When Twitter acquired BackType, it released Storm as an open source project. Storm entered Apache incubator status in September 2013 and graduated to top-level status in September 2014.

Storm applications express data transformations as a directed acyclic graph, where the vertices or nodes represent data sources or transformations and the edges represent streams of data flowing from one vertex to the next. Nodes that represent streaming data sources are called *spouts*; nodes that process one or more input streams and produce one or more output streams are called *bolts*; the complete network of spouts, bolts, and streams is called a *topology*.

Storm's messaging interface is sufficiently flexible that it can be integrated with any source of streaming source. Well-documented queue integrations currently include Kestrel, RabbitMQ, Kafka, JMS, and Amazon Kinesis.

Topologies are inherently parallel and run across a cluster of machines.[53] Different parts of the topology can be scaled individually by manipulating their parallelism. Storm's inherent parallelism means it can process very high throughputs of messages with very low latency.

Storm is fault-tolerant: if one or more processes fails, Storm will automatically restart it. If the process fails repeatedly, Storm will reroute it to another machine and restart it there.

Under its standard configuration, Storm guarantees "at-least-once" processing, which ensures that every incoming message will be processed. For exactly-once processing—a key requirement in financial systems—Storm supports an overlay application called Trident.

Once started, Storm applications run indefinitely and process data in low latency as it arrives.

Some of the ways that organizations use[54] Storm include:

- Groupon uses Storm to build low-latency data integration systems that can analyze, clean, normalize, and resolve large quantities of data.

---

[53]https://storm.apache.org/about/scalable.html
[54]http://storm.apache.org/documentation/Powered-By.html

- Spotify uses Storm for low-latency music recommendations, monitoring, analytics, and ad targeting.

- Cerner uses Storm to process clinical data in low latency.

- Taobao uses Storm to extract useful information from machine logs.

- TheLadders uses Storm to send hiring alerts; when a recruiter posts a job, Storm processes the event and aggregates job seekers who match the required profile.

Other applications include synchronizing contact lists, systems monitoring, trending topic detection, sentiment analysis of social media, security monitoring, and many others.

Hortonworks and MapR distribute and support Storm. Code contributions[55] have accelerated markedly since mid-2015.

Streaming Analytics in Action

Organizations currently use streaming analytics for a variety of use cases, including risk management, telco operations, for basic science, and for medical research, among others. Here are seven examples.

Credit Card Fraud.[56] Shopify offers an ecommerce platform as a service for online stores and retail point-of-sale systems. As of late 2015, the company reports that its platform supports more than 200,000 merchants and $12 billion in gross sales at a rate of 14,000 events per second. Processing credit card transactions is risky, and the company has just seven analysts to investigate possible fraud. Shopify processes transactions from Apache Kafka in Spark Streaming, filtering the riskiest transactions and routing them to case management software for investigation.

Credit Card Operations.[57] To detect unusual behavior, ING clusters venues (stores) according to usage patterns, then monitors the stream of transactions with Spark Streaming to identify venues whose cluster assignment changes. Analysts investigate anomalies to determine causes of the unusual behavior.

[55]https://www.openhub.net/p/apache-storm
[56]http://www.slideshare.net/SparkSummit/realtime-risk-management-using-kafka-python-and-spark-streaming-by-nick-evans
[57]http://www.slideshare.net/SparkSummit/realtime-anomoly-detection-with-spark-mlib-akka-and-cassandra-by-natalino-busa

Customer Experience Management.[58] Capital One, a leading U.S. consumer and commercial banking institution, has an overall technology strategy that seeks to shift data processing from batch operations to stream processing. In its digital operations, the bank monitors customer activity in low latency to proactively detect and resolve issues, prevent systems issues from adversely impacting the customer, and to enable a flawless digital experience. Capital One previously used expensive proprietary tools that offered limited capabilities for low-latency advanced analytics. The bank developed a new system that uses Apache Flink to process events from Apache Kafka and generate low-latency alerts, time-window aggregates, and other operations. Flink also provides the bank with the ability to perform advanced windowing, event correlation, fraud detection, event clustering, anomaly detection, and user session analysis.

Telco Network Monitoring.[59] Bouygues Telecom is one of the largest communications providers in France, with more than 11 million mobile subscribers, 2.5 million fixed line customers, and revenue of more than 5 billion euros. Bouygues' Logged User Experience (LUX) system captures massive quantities of log data from network equipment and generates low-latency diagnostics and alerts. Flink collects log events from Apache Kafka at an average rate of 20,000 events per second, transforms the raw data into a usable and enriched format, and returns it to Kafka for additional handling. Flink also generates alarms if it detects failures exceeding a threshold.

SK Telecom, South Korea's largest wireless carrier, offers[60] another example. SKT captures 250 terabytes of network logs per day, which it loads into a Hadoop cluster that now has more than 1,400 nodes. For low-latency analytics, the company uses Spark Streaming to capture events from Kafka and produce low-latency metrics of network utilization, quality, and fault analysis.

Neuroscience.[61] The Freeman Lab at Howard Hughes Medical Institute's Janelia Research Campus explores neural computation in behaving animals at the scale of large populations and entire brains. To facilitate this work, the lab has developed three open source packages: thunder[62], a large-scale imaging and time series analysis tool that runs on Apache Spark; lightning[63], a tool that produces web visualizations; and binder[64], software for reproducible computing with Jupyter

---

[58]http://www.slideshare.net/FlinkForward/flink-case-study-capital-one
[59]http://data-artisans.com/flink-at-bouygues-html/
[60]http://www.slideshare.net/SparkSummit/big-telco-yousun-jeong
[61]http://www.jeremyfreeman.net/share/talks/spark5/#/
[62]http://thunder-project.org/
[63]http://lightning-viz.org/
[64]http://mybinder.org/

and Kubernetes. Zebrafish brains have about 100,000 neurons (compared to human brains, which have 100 billion neurons). A scanning electron microscope working with a zebrafish brain produces two terabytes of data an hour. To measure response to various stimuli, researchers at the lab use low-latency analytics to capture key metrics, perform dimension reduction, clustering, and regression analysis, with results piped into visualization tools.

Medical Research.[65] In partnership with Intel and a consortium of hospitals, rehabilitation clinics, and clinical trial providers, the Michael J. Fox Foundation conducts research into Parkinson's Disease. A key challenge for researchers is a lack of objective measurements of the physical symptoms of the disease, including tremors; these symptoms progress slowly, and changes are hard to detect. With wearable devices and a stack of open source components that includes Apache Kafka and Apache Spark Streaming, scientists can monitor patient activity, symptoms, and sleep patterns, and they can correlate these with medication intake.

Streaming Economics

How important is low latency in analytics? In this and previous chapters, we identified two proven examples of disruptive analytics where reducing latency was a critical success factor:

- In financial markets, trading algorithms operate in a Darwinistic world, where microseconds matter.

- In fraud detection, credit card issuers must detect fraud within a narrow window of time or absorb the loss.

We also presented examples of streaming analytics at work in a number of fields: credit card fraud detection, credit card operations, customer interaction management, medical research, and telco operations. Most of the examples are operational, not strategic, and they appear to be disruptive. None of the applications is disruptive, as defined in Chapter One, with the possible exception of the medical research example, which could have a profound impact on research into Parkinson's Disease.

Overall, the economic impact of streaming analytics is very small outside of some clear-cut use cases. At the beginning of this chapter, we noted that spending on software for streaming analytics is only about 1% of total spending on software for business analytics. Spending on streaming analytics is growing faster than for the category as a whole, but even at the most optimistic projection, it's not likely to account for more than 2% of the category in 2020. That is a niche market.

---

[65]http://www.slideshare.net/SparkSummit/enable-breakthrough-in-parkinson-disease-research-ido-karavany

TIBCO, with its focus on a broad range of tools to reduce latency, was able to build a billion dollar business over a decade. Other players in the market were not so successful:

- When Sybase acquired Aleri in 2010, it acquired "certain assets" of the company, and not the company as a whole—which implies that Aleri was failing and no longer a going concern.

- Progress Software acquired APAMA for $25 million in 2005 and sold it in 2013 for $44 million, a modest gain. But revenue from the product line declined by 61% in the two years prior to the sale. Keep in mind that APAMA was and is the market share leader in the category.

- When TIBCO acquired StreamBase in 2013, the purchase price of $49 million barely covered StreamBase's total capitalization of $44 million.

In other words, among the companies that entered the streaming analytics market prior to 2005, there were no unicorns.

Even TIBCO, when it sold[66] itself to a private equity buyer in 2014, did so at a valuation of about four times revenue, a valuation appropriate for a mature company with limited growth prospects. At the time of the transaction, TIBCO's revenue had declined for more than a year, a problem TIBCO's CEO attributed to a transition to subscription pricing[67]—a sure sign of disruption.

The real-time analytics market also witnessed one of the greatest examples of hype-driven bubbles in the case of CRM provider E.piphany Inc. E.piphany, founded in 1996, went public in September 1999, trading at $16. Two months later, the stock traded at $80.

In November 1999, E.piphany announced that it was acquiring real-time customer interaction management company Rightpoint for $392 million. Investors went wild, bidding the stock up to $158.

Four months later, with the stock trading at $317, E.piphany announced that it was acquiring[68] Octane, another real-time customer interaction management provider, for $3.2 billion, 91 times projected revenue.

---

[66]http://www.tibco.com/company/news/releases/2014/tibco-to-be-acquired-by-vista-equity-partners-for-24-00-per-share-in-cash
[67]http://www.it-director.com/blogs/banks-statement/2014/9/is-tibco-a-worrying-sign-of-a-different-malaise/
[68]http://www.internetnews.com/bus-news/article.php/321151/Epiphany+Buys+Octane+Software+for+32+Billion.htm

E.piphany's revenue peaked in 2000, then declined precipitously. By 2004, the stock traded at \$3.50. SSA Global, an ERP vendor, acquired[69] the company for \$4.20 a share in 2005.

The business success of companies in the streaming and real-time analytics market is only a proxy measure of a technology's impact on the analytics value chain. However, when most companies in the category struggle to drive value, the implication is that there is little value to be driven.

The past, of course, does not necessarily foretell the future. TIBCO's struggles beginning around 2012 imply rapid adoption of cloud and open source streaming technologies, whose business model may make streaming attractive for new use cases.

Streaming advocates are excited about its potential for the Internet of Things (IoT). The Internet of Things (IoT) is the network of objects and devices, including vehicles, machines, buildings, and other devices embedded with sensors and connected to other devices. Examples include smart grids, smart vehicles, smart homes, and so forth. Connected devices generate huge volumes of streaming data.

We note that IoT is at the peak[70] of its hype cycle as of this writing.

Managers must distinguish between the costs of streaming data processing, on the one hand, and the benefits of reduced latency.

For human BI and interactive query workloads, latency measured in seconds is more than adequate; few human analysts tracking events through a BI system can benefit from lower latency than that.[71] Moreover, it is doubtful that any company ever lost money because junior program analysts had to wait a few minutes to view the results of a creative test.

The real potential for streaming analytics is in automated processes, where streaming engines can make decisions in a consistent manner, with much higher throughput and lower latency than humans can possibly deliver. Automation with streaming analytics will yield the highest economic benefits when applied to repeatable operations that are highly labor intensive, in operations where humans perform poorly, or in operations that are impossible for humans to perform.

---

[69]http://searchcrm.techtarget.com/news/1112932/SSA-Global-buys-into-CRM-with-Epiphany-acquisition

[70]http://www.gartner.com/newsroom/id/3114217

[71]Usability researchers report varying maximum acceptable response times; context matters. For example, see https://www.nngroup.com/articles/powers-of-10-time-scales-in-ux/ for a discussion of standards in customer-facing applications. For internal applications, standards are lower. See https://www.microstrategy.com/it/press-releases/microstrategy-introduces-new-high-performance-standards-for-business-intelligence.

Analytics in the Cloud

The Disruptive Power of Elastic Computing

Several years ago, SAS, one of the leading commercial business analytics software vendors, held an annual sales meeting for financial services account executives. Jim Goodnight, founder and CEO, spoke to the assembled sellers and led a Q&A session.

"What is our strategy for cloud computing?" asked one of the sales reps.

"There's only one thing you need to know about cloud computing," drawled Goodnight. "It's all BS." He compared cloud computing to mainframe service bureaus in the 1970s, which typically offered metered pricing based on usage. Goodnight repeated the comparison in a recent interview with The Wall Street Journal.[1]

Of course, everyone understands that modern cloud computing isn't the same as mainframe timesharing, any more than a 2016 Ford Focus is the same as a 1976 Ford Pinto because they both have four wheels. Goodnight is right, though, to point out some similar principles—sharing IT resources across multiple users and metered pricing.

---

[1]http://blogs.wsj.com/cio/2016/03/08/sas-institute-ceo-the-godfather-of-analytics-sees-future-in-the-past/

© Thomas W. Dinsmore 2016
T. W. Dinsmore, Disruptive Analytics, DOI 10.1007/978-1-4842-1311-7_7

Whether cloud computing is radically new or "all BS," as Goodnight suggests, it's clear that cloud is eating the IT world. A leading analyst forecasts[2] that cloud data center workloads will *triple* from 2013 to 2018; moreover, 78% of *all* data center workloads will be in the cloud by 2018.

We note, too, that shortly after Goodnight's original remarks, SAS announced[3] plans to invest $70 million in a cloud computing data center.

Cloud Computing Fundamentals

The National Institute of Standards and Technology (NIST) defines *cloud computing* as:

> ...*a model for enabling ubiquitous, convenient, on-demand network access to a shared pool of configurable computing resources...that can be rapidly provisioned and released with minimal management effort or service provider interaction.*[4]

Cloud computing has five essential characteristics:

- **On-demand self-service**: Users can provision the computing resources they need, such as server time and storage space, *without help from a systems administrator.*

- **Broad network access**: Users can access computing services over a network from diverse connected platforms, including mobile phones, tablets, laptops, and workstations.

- **Resource pooling**: The service provider pools computing resources to serve multiple consumers, with physical resources assigned and reassigned on demand. The user has no control or knowledge of the specific physical resources assigned.

- **Rapid elasticity**: The consumer can acquire and release resources on demand; to the consumer, computing resources appear to be unlimited, and can be acquired and released in any quantity at any time.

---

[2]http://www.cisco.com/c/en/us/solutions/collateral/service-provider/global-cloud-index-gci/Cloud_Index_White_Paper.pdf
[3]http://www.businesswire.com/news/home/20090319005110/en/SAS-Build-70-Million-Cloud-Computing-Facility
[4]http://csrc.nist.gov/publications/nistpubs/800-145/SP800-145.pdf

- **Measured service**: The cloud provider measures computing services on appropriate dimensions, such as server time or storage volume and time. Resource usage is monitored and reported in a manner that is transparent to the user and provider.

The NIST defines four cloud deployment models: private, community, public, and hybrid.

- **Public cloud**: The cloud provider owns and maintains a cloud platform and offers services to members of the general public.

- **Private cloud**: The cloud provider provisions a cloud for its own exclusive use. For example, a company builds and maintains a cloud computing platform and offers services to its own employees and contractors. The cloud may be owned and managed by the organization, a third party, or a combination thereof, and may reside on- or off-premises.

- A virtual private cloud (VPC) uses a software framework to provide the equivalent of a private cloud on public cloud infrastructure. VPCs are well-suited to the needs of small and medium businesses (SMBs).

- **Community cloud**: Groups of organizations, such as trade associations or affinity groups, provision a community cloud for the exclusive use of member organizations. Like private clouds, community clouds may be owned by one or more of the affiliated organizations, a third party, or a combination thereof, and may reside on- or off-premises.

- **Hybrid cloud**: An organization combines infrastructure from multiple deployment models (private, community, or public). For example, a company runs its own private cloud, which it supplements with public cloud during periods of peak workload. A software framework unifies the hybrid cloud, so a user does not know whether a job runs on the private or public infrastructure. A full 70% of organizations surveyed by IDC report a hybrid approach to cloud computing.

Consultant IDC projects[5] worldwide public cloud revenue to grow from nearly $70 billion in 2015 to more than $141 billion in 2019. In 2016, Amazon Web Services dominates the market; Microsoft ranks second

---

[5]https://www.idc.com/getdoc.jsp?containerId=prUS40960516

and is growing rapidly. Other vendors include Google, VMware, IBM, DigitalOcean, and Oracle.[6]

IDC estimates[7] total spending on private cloud infrastructure of $12.1 billion in 2015; consultant Wikibon estimates[8] a much lower level of $7 billion. (Private cloud is harder to measure than public cloud.) Leading suppliers in the market include Hewlett Packard Enterprise (HPE), Oracle, VMware, EMC, and IBM. The private cloud market is much more fragmented than the public cloud market, and the top ten vendors control only about 45% of spending.

One consultant projects[9] total spending on VPCs to exceed $40 billion by 2022. Community clouds are a much smaller presence in the market, with total spending projected[10] to reach $3.1 billion by 2020.

Within the NIST's broad deployment models, there are three distinct services models for cloud computing:

- **Infrastructure-as-a-Service (IaaS)**: The most basic cloud service: the provider offers fundamental processing, storage, and networking services together with a virtual resource manager. The end user installs and maintains the operating systems and application software.

- **Platform-as-a-Service (PaaS)**: The provider offers a complete computing platform, including operating system, programming languages, databases, and web server. The end user develops or implements applications on the computing platform, but the cloud provider is responsible for installation and maintenance of the supported components.

- **Software-as-a-Service (SaaS)**: The end user has direct access to application software. There are two distinct SaaS models: direct selling and marketplace selling. Under a *direct selling* model, the software vendor handles software application and maintenance and either hosts

---

[6]http://assets.rightscale.com/uploads/pdfs/RightScale-2016-State-of-the-Cloud-Report.pdf

[7]http://www.idc.com/getdoc.jsp?containerId=prUS25946315

[8]http://wikibon.com/public-cloud-iaas-is-3-5x-the-size-of-true-private-cloud-adoption/

[9]https://www.infoholicresearch.com/press-release/end-decade-small-medium-enterprises-smes-will-dominate-virtual-private-cloud-market-infoholic-research/

[10]http://www.strategyr.com/MarketResearch/Community_Cloud_Market_Trends.asp

the software on its own infrastructure or contracts with a cloud provider for IaaS or PaaS services. Under the *marketplace selling* model, the cloud provider creates a platform where end users can shop and select software from many vendors; the cloud provider handles installation and maintenance of the application software.

Consultancy Technology Business Research (TBR) estimates that SaaS currently accounts for about 60% of the public cloud market; Salesforce is the clear leader in SaaS. The PaaS market is relatively small, accounting for about 10% of the 2016 market; TBR projects a 17% growth rate through 2020. Microsoft leads in PaaS, followed by IBM, Salesforce, and Google.

As a general rule, IaaS services target developers and IT organizations, who configure and manage the platform for the benefit of their end users. PaaS services target developers and power users willing and able to perform basic software configuration tasks. SaaS services target business users.

The distinction among the service models tends to be blurred in practice. Amazon Web Services, for example, is generally considered to be an IaaS provider; but AWS offers PaaS and even SaaS services. As cloud providers seek to expand the reach and profitability of their offerings, we can expect they will seek to move up the value chain, offering higher level services and solutions.

Cloud computing offer benefits similar to managed software hosting, but the business model and technologies differ. Software hosting dates back at least to the mainframe service bureau model of the 1960s; in the 1990s, service bureaus evolved into application service providers (ASPs) with a more diverse technology stack.

Unlike cloud service providers, ASPs host software and applications for customers under contract. The customer licenses software from a software vendor; instead of deploying the software on-premises, however, the hosting provider implements and manages the software under a long-term contract. The ASP generally does not pool IT infrastructure across customers; instead, the ASP builds the cost of dedicated infrastructure into contract costs. Contract terms generally call for fixed periodic payments, with a schedule of extra services billed as used.

ASPs tend to operate as partners of large software vendors. The software vendor's marketing muscle helps the ASP's business development effort, and the ASP's hosted model helps the software vendor sell to companies with limited IT skills and capacity. Many large software vendors, like SAS, operate their own in-house ASP operations.

Cloud computing rests on three innovations:

Virtualization. Hardware virtualization is a process that creates one or more simulated computing environments ("virtual machines" or "instances") from a single physical machine. Virtualization improves IT efficiency, since individual applications rarely make full use of modern computer hardware.

The primary goal of virtualization is to manage workloads by making computing more scalable. Virtualization is not new, but has progressively evolved for decades. Today it can be applied to a wide range of system layers, including operating system, applications, workspaces, services, and hardware components (including memory, storage, file systems, and networks).

Virtualization provides other benefits in addition to improved hardware utilization. For example, IT organizations can more easily administer security and access control on virtual instances, and they can quickly redeploy an instance from one physical machine to another when necessary for maintenance or failover.

Service-Oriented Architecture (SOA). Service-oriented architecture is a computer design architecture in which application components provide services to other components through a communications protocol. Services are self-contained, loosely coupled representations of repeatable business activities. Breaking complex applications down into self-contained services simplifies development, maintenance, distribution, and integration.

Autonomic Computing. In 2001, IBM coined[11] the term *Autonomic Computing* to refer to an approach to distributed computing that seeks to build self-regulating systems with autonomic components. Autonomic components can configure, heal, optimize, and protect themselves.

The technologies that enable the cloud are not new. The extreme growth of companies like Amazon, Google, and Facebook and the commensurate demands on their IT infrastructure produced a perfect storm of innovation and skills in virtualization, provisioning, energy consumption, security, and other disciplines required to run a modern data center.

The Business Case for Cloud

In this section, we discuss the economic benefits of cloud and the use cases that drive firms to use the cloud, together with concerns about data movement and security.

Cloud Economics

Is cloud computing more expensive or less expensive than on-premises computing? The answer is a matter of some controversy. One reason the problem is challenging is that many firms do not accurately measure the total costs of their IT infrastructure; they measure "hard costs" of equipment and purchased software, but fail to measure "soft costs" of IT personnel. Cloud computing costs,

---

11http://ieeexplore.ieee.org/xpl/articleDetails.jsp?arnumber=1160055

on the other hand, are tangible; at the end of the billing period the vendor sends an invoice, so costs are clear. This tends to create a bias against cloud in some organizations, especially so if IT executives see cloud computing as a threat.

Cloud computing vendors enjoy a number of advantages compared to on-premises IT organizations.

Economies of Scale. Vendors like Amazon Web Services and Google purchase hardware in huge quantities. They have highly capable purchasing organizations who negotiate hard bargains with hardware vendors.

Economies of Skill. Public vendors hire the best and brightest people to manage their data centers. They are extremely competent managers with deep experience running massive worldwide networks. In every aspect of data center management, from virtualization to energy consumption, they are simply better at their jobs.

Utilization Economies. By pooling resources and spreading them over a large worldwide user base, public cloud vendors achieve a much higher level of utilization for IT infrastructure than individual firms are able to accomplish. This means cloud vendors can charge a lower unit price for computing resources.

As a result of these economies, combined with aggressive competition among the leading cloud vendors, public cloud charges declined[12] by double digit percentages each year prior to 2016.

Of course, these economies do not apply across the board. Some large organizations have the scale and the skill to deliver unit computing costs that match or beat cloud pricing. Organizations with accurate total cost of ownership (TCO) metrics can confidently assert that it makes good economic sense to keep computing on-premises.

But there a number of logical use cases for the cloud even for those firms with lower unit costs of computing.

Predictable Peak Workload. Many organizations have workloads with predictable peaks that are significantly higher than the base workload. For example, retailers process much higher transaction volumes during the peak Christmas season, and most organizations have high month-end processing for accounting and financial applications. If an organization builds IT infrastructure to meet its peak workloads, that infrastructure will be underutilized most of the time. A hybrid approach that uses cloud for peak workloads and on-premises systems for base workload is the optimal approach for these organizations.

---

[12]https://451research.com/images/Marketing/press_releases/03.01.16_CPI_North_America_PR_FINAL.pdf

Unexpected Peak Workload. A similar calculus applies for organizations with an unexpected surge in IT workload. Rather than rushing to add more infrastructure to support a sudden and unexpected increase in demand, it makes sense for the organization to use the cloud to support the incremental workload, at least until a root-cause analysis is complete.

Variable Cost Businesses. Many businesses in the services industry are inherently variable cost businesses; they are thinly capitalized and operate under a model where costs are charged back to clients and projects. Consulting firms, advertising agencies, marketing services providers, analytical boutiques, and other similar firms generally prefer to avoid investing in overhead of any kind, including IT infrastructure. Firms in this category may choose to rely exclusively on the cloud to support project work.

Time to Value. In many cases, executives are less interested in costs and more interested in time to value. This is often true for rapidly growing businesses; it may also be true for firms whose IT organizations are operating at or near capacity. For these firms, cloud offers immediate capacity and an ability to scale up quickly. Speed and time to value are key value propositions stressed by SaaS vendors like Salesforce, which cater to business needs for rapid capability.

Business Unit Autonomy. Business units sometimes choose to work independently of the IT organization. This may be due to real or perceived shortcomings of IT's ability to support the business effectively, political conflict between executives, or competition for scarce IT resources from other business units. In any case, turning to the cloud offers the business unit executive an opportunity to control the IT resources needed for a mission-critical initiative.

Pilots and POCs. Certain business analytics use cases are very attractive for the cloud. Among these are pilot projects and proof of concepts (POCs): projects designed to determine the viability of a specific solution. Cloud computing is attractive for these projects because an organization can quickly provision a temporary environment without the risk of a sunk infrastructure cost. If the project proves viable, the organization can keep the application in the cloud or port it to an on-premises platform. If the pilot or POC is not successful, the organization simply shuts it down.

Ad Hoc Analytics. Strategic ad hoc projects are also an attractive use case for cloud. Many analyses performed for top executives are not repeatable; the analysis will be performed once and never again. Conventional approaches to data warehousing do not apply to ad hoc analysis, which is surprisingly prevalent in most enterprises. The cloud enables analysts to quickly create a temporary datastore with as much storage and computing power as needed.

Model Training. The model training phase of machine learning is particularly well suited to elastic computing in the cloud, for several reasons:

- Model training is project-oriented rather than a recurring production activity. (Model scoring, on the other hand, is production oriented.)

- Machine learning algorithms require a lot of computing power, but for a short time only.

- Machine learning projects often support development projects performed in advance of IT infrastructure investments.

- In areas such as marketing, projects are often started with little lead time and require rapid provisioning.

- To deliver machine learning projects, organizations frequently engage analytic service providers (ASPs), who must account for infrastructure costs.

Many enterprises outsource ad hoc analysis to consulting firms and analytic service providers. Cloud computing is especially important to these firms, because the cloud's low cost and measured service enables them to explicitly match computing costs to client projects and to expand quickly without capital investment.

Data Movement

Data movement is always a concern when working with Big Data. A process that takes minutes with a small data set can take hours or days when we measure data in petabytes. In the Big Data era, minimizing data movement is a key governing principle.

Nevertheless, in business analytics at least some data movement is inevitable. IT organizations rarely permit production systems that serve as data sources to be used for analytics; in any case, these systems generally lack the necessary tooling. Hence, in most cases each piece of data will be copied and moved at least once, when it is transferred from a source system to an analytic repository.

Moving data to and from the cloud poses even greater concerns than moving data internally. Public networks can be a bottleneck, and security concerns dictate an encrypt/decrypt operation at either end to avoid a data breach. To mitigate the problem, cloud providers and their alliance partners offer a number of products and services.

Dedicated Physical Connections. Cloud providers offer customers the ability to connect directly to the provider's data center through a dedicated private network connection. Services like this are a good choice for organizations seeking to move regular updates to a cloud-based environment for analytics.

Data Transfer Accelerators. The leading cloud providers operate global networks of data centers and will work with customers to optimize data transfer. For example, an organization with operations in multiple countries may be able to minimize data transfer costs by transferring data locally in each country, then consolidating the data within the cloud provider's network.

Portable Storage Appliances. Secure transportable storage devices are now available to store up to 80 terabytes (TB); larger files can be split across multiple devices. Organizations seeking to move data to or from the cloud copy data to the secure device, then send to the destination data center by overnight delivery. This method is excellent in scenarios where a mass of data will be moved all at once, as in a database migration or in the early stages of an analysis project.

Storage Gateways. For organizations seeking to implement a hybrid architecture that mixes on-premises and cloud platforms, a storage gateway may be the right solution. Storage gateways (from firms such as Aspera (IBM), CloudBerry, NetApp, and Zerto) reside on customer premises and broker between on-premises and cloud storage, maintaining a catalogue of data and its location. Storage gateways handle compression and encryption.

Database Migration Services. When the organization seeks to migrate or update data from an on-premises relational database to a database in the cloud, it is highly desirable to maintain the structure and metadata. Extract and reload operations take time, because the database administrator must map the structure of the source database to the target database; they are also subject to human error in the mapping process. Tools from vendors like Attunity enable the organization to copy data directly from database to database, on-premises, in the cloud, or both.

Of course, cloud platforms are the logical site for business analytics when the source data is already in the cloud and does not need to be transferred. Some businesses have built their operations around cloud computing; for these organizations, data movement to and from the cloud is not an issue.

Security

Vendors of on-premises software cultivate the perception that the cloud is less secure than on-premises data management. In many organizations, there are executives who believe that their data is most secure when it remains on-premises; they oppose moving data to the cloud.

The data suggest otherwise. On-premises facilities suffered[13] ten out of ten of the worst data breaches in 2015. A cybersecurity report issued by the Association of Corporate Counsel reveals[14] that employee error is the leading cause of data breaches; a separate analysis performed by privacy and data protection specialists Baker & Hostetler LLP found that employee negligence is the biggest cause of breaches.

Effective security stresses good management policies and practices rather than the physical location of the data. The leading public cloud vendors are very good at managing data center security; they go to great lengths to certify compliance with security standards published by ISO and NIST and as required under HIPAA and other governing legislation. There are no known security breaches[15] recorded by the leading cloud providers in a decade of service.

Executives surveyed[16] by consultant Technology Business Research (TBR) cited security as their top consideration in cloud decision making. However, respondents also said they believe having their data stored and managed by an expert third party improves overall security. In the same survey, respondents indicated that security is the primary consideration favoring a private cloud; on all other dimensions, respondents rated a public cloud equal to or better than a private cloud.

Personally Identifiable Information (PII) is especially sensitive data, since its unauthorized disclosure directly impacts consumer privacy and exposes the organization to serious consequences. The definition of PII is surprisingly broad, because bad actors can combine PII from multiple sources to create a detailed profile of the prospective victim. A person's Social Security Number (SSN) is obviously PII; but fraudsters can predict[17] a victim's SSN from date of birth and place of birth.

Personally Identifiable Information (PII) is information that can be used to identify, contact, or locate an individual. In the United States, the National Institute of Standards and Technology publishes standards for what constitutes PII and how to manage it.

---

[13]http://www.crn.com/slide-shows/security/300077563/the-10-biggest-data-breaches-of-2015-so-far.htm
[14]http://www.acc.com/aboutacc/newsroom/pressreleases/accfoundationstateofcybersecurityreportrelease.cfm
[15]http://www.cnet.com/news/cloud-computing-security-forecast-clear-skies/
[16]http://www.slideshare.net/TBR_Market_Insight/soaring-toward-113b-tbr-projects-key-trends-in-cloud
[17]http://www.pnas.org/content/106/27/10975.full

While operational systems must capture and retain PII, it is often not needed in business analytics. There are exceptions to that generalization; an analyst may want to apply machine learning techniques to customer surnames to identify ethnicity or geocode a customer's address to perform spatial analysis. When it is essential to work with PII, good working methods minimize the security risk; we discuss these in Chapter Ten.

Analytics in the Public Cloud

In this section we examine cloud services pertinent to business analytics from Amazon Web Services, Microsoft, and Google. We focus on managed services for storage, compute, Hadoop, relational databases, business intelligence, and machine learning; all three vendors offer many other services, which could be pertinent for some projects. The intent is to cover the most widely used services.

We show pricing information for reference and comparison, but the reader should bear in mind that cloud vendors can and do change prices frequently.

While we focus on managed services, all three vendors offer basic compute and storage services, which enable an organization to implement *any* licensed software in the cloud.

Amazon Web Services

Amazon Web Services (AWS) offers more than 50 managed services. Of these, the services most pertinent to business analytics are storage services, compute services, Hadoop services, database services, business intelligence services, and machine learning services.

Storage Services

For any cloud platform, data storage is the most fundamental service, for two reasons. First, to serve as an initial staging area for data when it is first uploaded to the cloud; second, to serve as persistent storage after an application finishes processing.

In AWS, most computing instances include some local storage, which is available as long as the instance is available. However, anything saved in the local storage will be lost when the user's lease on the compute instance expires. Thus, it makes sense to separate long-term file storage from compute instances, which users may want to rent briefly, then release.

Amazon Simple Storage Service (S3) was the first service offered by AWS, in March 2006. It is a low-cost scalable storage service that stores computer files as large as five terabytes. Files can be in any format.

Applications interact with S3 through popular web services interfaces, such as REST, SOAP, and BitTorrent; this enables S3 to serve as the back-end storage for web applications. According to AWS, the S3 service uses the same technology that Amazon.com uses for its ecommerce platform.[18]

AWS offers three different storage classes at different price points: Standard, Infrequent Access, and Glacier.

- Standard storage offers immediate access with no minimum file size, no minimum storage duration, and no retrieval fees.

- Infrequent Access storage offers immediate access with a retrieval fee, a minimum file size, and a 30-day minimum storage. Monthly costs are lower as long as the user does not retrieve files frequently.

- Glacier storage offers access with up to four hours' latency with a retrieval fee and a 90-day minimum storage.

AWS charges monthly fees for S3 usage by the gigabyte (GB). Pricing depends on the storage class and total storage used. For example, in the US East region, the price per GB per month for Standard storage up to 1 terabyte (TB) is $0.03; for Infrequent Access storage, the price is $0.0125 per GB per month; for Glacier storage, $0.007. Prices for all three services decline with volume. As is the case for all AWS services, prices may vary by region.[19]

Compute Services

Amazon Elastic Compute Cloud (EC2) is one of the earliest services offered by AWS and remains a foundation of its cloud platform. EC2 enables users to rent virtual machines to provision their own applications, paying for machine time by the hour.

Users can start and end server sessions as needed, choosing from a wide range of instances, priced according to computing power. AWS's available server types change constantly, falling into five different categories:

- *General Purpose* instances provide a balance of compute, memory, and network resources. Some instances in this category are burstable, which means they offer a baseline capacity with the ability to "burst" temporarily above the baseline.

- *Compute Optimized* instances feature the highest performing processors (measured by CPU speed).

---

[18]http://aws.amazon.com/s3/
[19]All pricing is as of May 2016.

- *Memory Optimized* instances have larger amounts of Random Access Memory (RAM) per CPU.

- *GPU* instances include one or more NVIDIA GPUs.

- *Storage Optimized* instances include high I/O instances with SSD backed storage and dense storage instances with very large hard drives per CPU.

Within each category, AWS offers sizes ranging from small instances equivalent to a laptop computer to extra large instances with thousands of cores. Instances are preconfigured with an operating system image; users can choose from among Linux (Amazon, Red Hat, SUSE, or CentOS), Windows (with and without SQL Server), Debian, and other operating systems.

EC2 users can predefine virtual appliances, called Amazon Machine Images (AMIs). These consist of an operating system and any other software needed to run an application. AMIs make it cost effective to run complex software stacks on EC2, since the user need not use valuable instance time installing and configuring software.

AWS offers three main pricing models for EC2: on-demand, reserved, and spot pricing. Under on-demand pricing, the user pays by the hour with no commitment. Pricing varies by region and operating system; in April 2016, rates on Amazon Linux in the US East Region ranged from .0065 to $6.82 per hour.

Reserved instance pricing provides the user with a discount in return for a commitment to use the instance for a defined term of one to three years. The discount is significant. For example, for a large general purpose instance (M4), the user pays $1,173 in advance for a three year term, which is $.0446 per hour; by comparison, the on-demand hourly rate is $0.12 per hour, almost three times higher. (Pricing is on Amazon Linux, US East Region, April 2016.) Of course, that comparison is valid only if the user runs an application on the instance constantly for the entire term of the contract.

The difference between on-demand and reserved instance pricing is comparable to the difference between renting a hotel and renting an apartment. A traveler visiting a city only needs a room for a limited number of nights and is willing to pay a relatively high price per night for the short-term stay. On other nights, the hotel rents the room to other travelers. On the other hand, a person residing in a city needs a place to stay all of the time and is willing to make a fixed commitment in return for exclusive use of the apartment for the term of the lease.

The Spot pricing model works like an auction market. Users bid for unused Amazon EC2 capacity; instances are charged the Spot Price, set by AWS and fluctuating with supply and demand.

Hadoop Services

In Chapter Four we covered the Hadoop ecosystem. Amazon Elastic MapReduce (EMR) is AWS's managed service offering for Hadoop. AWS first offered EMR in April, 2009.

EMR is elastic in two respects. Users can deploy multiple clusters without limit; clusters can be configured differently and use different instance types while still sharing storage. Users can also resize existing clusters, adding or dropping nodes even while a job is running in the cluster.

AWS offers EMR users the ability to work with different file systems, including S3, HDFS (the "native" Hadoop file system), Amazon DynamoDB (a NoSQL database service), Amazon RDS, and Amazon Redshift. EMR includes commonly used components from the Hadoop ecosystem, including HBase, Hive, Hue, Impala, Pig, Presto, Spark, and Zeppelin.

AWS prices EMR by the instance hour, one instance per cluster regardless of the number of nodes in the cluster. Pricing varies according to the type of instance used for the cluster. Users also pay EC2 costs. Hence, the total charge for a 10-node EMR cluster on large general purpose instances will be the charge for the 10 EC2 instances plus the charge for the EMR instance.

Database Services

Amazon Relational Database Service (RDS) is a distributed database service first offered in October 2009. As of April 2016, AWS offers managed services for MySQL; Oracle Database; Microsoft SQL Server; PostgreSQL; MariaDB; and Amazon Aurora (a high-availability version of MySQL).

Amazon Redshift is a petabyte-scale columnar data warehouse based on technology licensed from data warehouse vendor Actian. Redshift is suitable for SQL analysis on very large volumes of data.

In theory, users can set up their own databases on leased EC2 instances. The AWS services relieve the user of the need to install, configure, provision, and patch the database software, and it simplifies the process of scaling up compute and storage. AWS also provides automated backup and database snapshots.

Business Intelligence Services

AWS users can set up any BI tool in EC2 and query databases in RDS or Redshift. In October 2015, AWS announced[20] a public preview of its own

---

[20]https://aws.amazon.com/blogs/aws/amazon-quicksight-fast-easy-to-use-business-intelligence-for-big-data-at-110th-the-cost-of-traditional-solutions/

business intelligence service branded as Amazon QuickSight. QuickSight offers users the ability to connect to data in Redshift, RDS, EMR, S3, and many other data sources, and it uses a distributed in-memory calculation engine to deliver fast interactive visualization. AWS expects to release the service to production in 2016, priced at a monthly fee per user.

Machine Learning Services

Amazon Machine Learning (ML) is a managed service that works with data stored in Amazon S3 files, Amazon Redshift, or MySQL databases in Amazon Relational Database Service. The service includes tools for data visualization, exploration, and transformation and a limited number of machine learning algorithms.

For prediction, the service supports APIs for batch or real-time scoring. The service does not support model import or export.

AWS prices the service at a set price per hour to analyze data and build models and to separate volume-based prices for batch and real-time prediction.

Marketplace

AWS also hosts a marketplace platform that enables software vendors and other sellers to offer their capabilities. Vendors offer software as Amazon Machine Images, as AWS CloudFormation Stacks, or as Software-as-a-Service.

AWS CloudFormation is a service that helps users model and set up AWS resources. It offers broader capability than Amazon Machine Images.

In most cases, vendors offer trial versions of their software free of license fees. Others charge hourly, monthly, or annual license fees, with AWS handling metering and billing. In still other cases, users must license the software separately through other channels. In all cases, users are responsible for AWS storage and compute charges.

Microsoft Azure

Microsoft has a unique approach to cloud computing that reflects its strength in on-premises computing and in enterprise desktop software. The backbone of Microsoft's cloud services is Microsoft Azure, a specialized operating system that manages computing and storage resources. (The Microsoft Azure brand applies to both the cloud platform as a whole and to the cloud operating system.)

Storage Services

Microsoft offers several different types of storage for different types of data:

- Azure Table Storage for structured data
- Azure Blob (Binary Large OBjects) Storage for documents, videos, backups, and other unstructured text or binary data
- Azure Queue Storage for messages
- Azure File Storage for Server Message Block (SMB) files

These storage types are available in four different data redundancy options. The least expensive option retains three copies of the data within a single data center; the most expensive option distributes six copies to two geographically separated data centers, with read access for high availability.

Each of the four storage types carries a separate rate card. Blob storage is the least expensive and file storage the most expensive.

Microsoft also offers a premium storage service based on solid state drive (SSD) storage for I/O intensive workloads.

Compute Services

Like AWS, Microsoft offers virtual computing platforms in a wide range of instance types and operating systems. Microsoft also distinguishes between Basic instances designed for development and test environments, and Standard instances for production environments.

Computing platforms are not limited to the Windows operating system. Microsoft Azure offers instances with Linux, Red Hat Linux, SUSE Linux, CentOS Linux, and Canonical Ubuntu. Not surprisingly, Microsoft offers instances preconfigured with Microsoft applications, like SQL Server and SharePoint.

Hadoop Services

Microsoft Azure HDInsight is a managed service for Hadoop based on Hortonworks' HDP distribution, with some changes to reflect Microsoft standards and architecture. HDInsight is elastic in a manner similar to AWS's EMR; users can add or drop clusters, or they can add or drop nodes to existing clusters.

HDInsight includes the analytical components that are bundled with the standard Hortonworks distribution: HBase, Hive, Pig, Phoenix, Spark, and Storm. For an extra charge, users can also access Microsoft R Server, a distributed machine learning engine with R bindings.

Microsoft prices HDInsight per node. Prices vary based on the type of compute instance used for each node.

Database Services

Microsoft Azure SQL Database is a managed Database-as-a-Service offering based on Microsoft SQL Server. The service is comparable to AWS Relational Database Service, but limited to a single core database product. Database functionality is a subset of Microsoft SQL Server functionality.

Pricing for the service is in three tiers at different service levels. The unit of measure is the Database Transaction Unit (DTU), a measure of a database instance's ability to process transactions.

For petabyte-scale SQL processing, Microsoft Azure offers SQL Data Warehouse a columnar database comparable to AWS Redshift. Unlike Redshift, SQL Data Warehouse decouples compute and storage, so users can shut down the query engine without losing stored data. SQL Data Warehouse also includes a capability to query non-relational sources, such as delimited files, ORC storage, HDFS and Azure Blob Storage.

Microsoft bills SQL Data Warehouse compute resources in Data Warehouse Units (DWU), a measure of query performance. Storage is billed separately at Azure Blob Storage rates.

Business Intelligence Services

Microsoft's popular BI tools, such as Excel and PowerBI, can readily use Azure data management services as a back end. Users deploy the tools locally and configure them to use the Azure data source; the application performs complex computations in the cloud and transfers a result set to the local user.

Machine Learning Services

Microsoft Azure Machine Learning is an offering from Microsoft that includes a browser-based client, machine learning algorithms in the cloud, APIs for Python and R and a marketplace for applications.

Azure Machine Learning Studio is an interactive drag-and-drop development environment enabling users to build, test, and deploy machine learning applications. The service works with data in a wide range of formats, including text files, Hive tables, SQL tables, R objects, and many others. Azure Machine Learning supports feature engineering and a wide range of algorithms for regression, classification, clustering, and anomaly detection. User can embed custom Python and R modules in a machine learning pipeline and deploy models as a web service.

Microsoft prices Azure Machine Learning at a flat monthly rate per seat, plus hourly charges to use Machine Learning Studio and the Machine Learning API.

Google Cloud Platform

Google Cloud Platform (GCP) is a relative latecomer to the public cloud market, first offering storage services to the public in 2010 and compute services in 2012. Since that time, however, Google has progressively added services and now offers a nearly complete platform for business analytics.

Google does not currently offer a managed service for BI and visualization. However, many such tools can connect to Google BigQuery (discussed later in this chapter) and use it as a data source.

Storage Services

Google Cloud Storage (GS) is a storage service comparable to the AWS S3 and Azure storage services. GS stores objects up to five terabytes.

Like AWS, GS offers storage in three options, priced according to availability. Standard storage, the most expensive, offers immediate access without retrieval fees. Durable Reduced Availability storage is slightly less expensive than Standard storage, with lower guaranteed availability. Nearline storage offers the lowest monthly charge per gigabyte, but Google charges users for each retrieval and guarantees availability at the lowest level.

Unlike Microsoft, Google does not distinguish among types of objects stored.

Compute Services

The Google Compute Engine (GCE) offers virtual machines ("instances")[21] from the same global infrastructure that runs Google's branded services, such as Gmail and YouTube.

GCE categorizes instances as predefined or custom. Predefined instances have preset virtualized hardware properties at a set price. There are four classes of predefined instances: Standard, Shared Core, High Memory, and High CPU. Standard instances range from 1 to 32 cores; High Memory and High CPU instances range from 2 to 32 cores.

Users specify custom instances to include an even number of virtual cores up to 16 or 32 (depending on the user's region), and from 0.9 gigabytes (GB) to 6.5 GB of memory per virtual core (in multiples of 256 megabytes). Pricing depends on the number of virtual CPUs and memory.

---

[21]Google uses the term "virtual machine" to refer to what AWS calls an "instance". For consistency, we use the term "instance".

At a steep discount, Google offers pre-emptible instances. Google warns the user 30 seconds in advance to permit graceful shutdown.

GCE offers instances with Debian, CentOS, CoreOS, SUSE Linux, Ubuntu, Red Hat Linux (RHEL), Free BSD, and Windows. There are extra charges for RHEL, SUSE, and Windows.

While AWS and Microsoft charge for compute services by the hour, Google charges by the minute, with a 10-minute minimum. Sustained use discounts apply to instances used for specified percentages of the billing month. Google charges the same rates in all regions.

Hadoop Services

Google Cloud Dataproc is a managed service for Hadoop and Spark. Google first offered the service in beta in September 2015 and released[22] it to general availability in February 2016.

Cloud Dataproc includes core Apache Hadoop (MapReduce, HDFS, and YARN), Spark, Hive, Pig, and connectors to other GCP services, including Cloud Storage, BigQuery (discussed later in this chapter), and BigTable (discussed later in this chapter), all deployed on Debian. Google integrates the components in an image, updating image versions with major or minor releases to reflect releases and patches for any of the components. Users may select older versions to create new clusters for up to 18 months after version release.

As with AWS EMR and Azure HDInsight, Cloud Dataproc is fully elastic. Users can add and drop clusters or resize them as necessary.

Google charges one cent per hour for each virtual CPU in the Cloud Dataproc cluster.[23]

Database Services

For relatively small relational database applications, Google offers Cloud SQL, a managed service featuring the MySQL database. The service is comparable to the AWS Relational Database Service and Microsoft Azure SQL Database. Google charges an hourly rate per database instance scaled according to the size of the instances upon which the database is deployed. Storage is an extra monthly charge, and there are network charges for egress (traffic leaving the instance).

---

[22]http://venturebeat.com/2016/02/22/google-launches-cloud-dataproc-service-out-of-beta/
[23]Pricing as of May 2016.

Google BigTable is a petabyte-scale high performance data management system. It is not a relational database, but a massively scalable hypertable or multi-dimensional datastore. BigTable supports APIs for HBase and Google's Go programming language, and a connector to Google Cloud Dataflow, Google's general programming framework.[24]

BigTable does not support SQL, and is therefore not comparable to AWS Redshift or Microsoft Azure SQL Data Warehouse. For scalable SQL, Google recommends Google BigQuery, an SQL engine that works directly with Google Storage. BigQuery is an implementation of Dremel,[25] a Google core technology.[26] Google first placed Dremel in production in 2006 and uses it today for many applications.

BigQuery uses columnar storage and a tree architecture for dispatching queries. (We discussed columnar serialization in Chapter One.) Columnar storage enables a very high data compression ratio and minimizes scan time for analytic queries. Google's query tree architecture enables BigQuery to distribute queries and collect results over thousands of machines.

Google charges users for storage (at a rate equivalent to Google Cloud Storage) and a standard rate of $5 per terabyte (TB) for queries; the first TB is free. Certain computationally intensive queries do not qualify for the standard rate; Google classifies these as high-compute queries and prices them individually. Google does not disclose the computing limit for standard pricing. If the query exceeds the threshold, Google informs the user and provides a cost estimate; the user must expressly opt in to run the query.

While BigQuery provides functionality that is similar in many respects to Redshift and SQL Data Warehouse, Google's pricing model is quite different. Users of the AWS and Microsoft services determine the computing resources to be used and pay for what they request. Google BigQuery users pay for actual query processing volume, while Google determines the computing resources used for the query.

Machine Learning Services

Google Cloud Machine Learning is a managed service for Deep Learning. Users define models with the TensorFlow framework released to open source by Google in 2015. (We discuss TensorFlow in Chapter Eight.)

Cloud Machine Learning integrates with Cloud Dataflow for preprocessing and works with data held in Google Cloud Storage, BigQuery, and other data sources.

---

[24]Google donated Cloud Dataflow to Apache, where it is incubating as Apache Beam.
[25]https://cloud.google.com/files/BigQueryTechnicalWP.pdf
[26]Dremel is also the foundation of Apache Drill, an open source SQL engine.

As of April 2016, the service is in Limited Preview, and Google has not released pricing.

Google Cloud Vision API is a managed service that supports label detection, optical character recognition, explicit content detection, facial detection, landmark detection, logo detection, and image properties. The first 1,000 units are free; above that level, Google charges a price per 1,000 units. A unit is one feature, or service, applied to one image.

Google Cloud Speech API uses Deep Learning to convert audio to text. The API recognizes more than 80 languages and dialects and works with data uploaded directly or stored in Google Cloud Storage. The service is in Limited Preview.

Google Translate API offers a simple programing interface for rapid translation of any text into one of more than 90 supported languages. The translation engine detects source text language when the language is not known in advance. Pricing is $20 per million characters.

Google also offers the Google Prediction API, a service that works with uploaded training data in CSV format. Users can elect to work with Google's "black box" training algorithm or select a technique from a library of hosted models. The service is free for the first six months up to defined usage limits. After the free period has expired, Google charges a base monthly fee per Google Cloud Platform Console project, plus separate fees for model training and prediction. Pricing does not include the cost of Google Cloud Storage services to hold the training data.

The Disruptive Power of the Cloud

The *technologies* that enable cloud computing aren't new. It is the cloud *business model*—elastic, "pay for what you use" computing—that is disrupting the technology industry and, by extension, the leading business analytics providers.

It's clear that cloud computing is disrupting the computer hardware industry. In 2014, the National Resources Defense Council commissioned a study of data center energy efficiency. The study found[27] that servers in the cloud operate at about 65% of capacity, while on-premises servers operate at 12% to 18% of capacity. On-premises servers are used less because the organizations that operate them build capacity to meet peak demand, so the servers are idle most of the time.

If businesses shift peak workloads to the cloud, they don't buy as many servers. This is already happening, and companies whose businesses depend on selling computer hardware to other businesses are struggling:

- IBM reports[28] a 22% decline in server systems sales.

---

[27]https://aws.amazon.com/blogs/aws/cloud-computing-server-utilization-the-environment/
[28]https://www-03.ibm.com/press/us/en/pressrelease/49554.wss

- Chipmaker Intel plans[29] to slash 12,000 jobs on disappointing sales.

- Storage vendor EMC reports[30] declining product sales.

Interestingly, while Intel's other businesses are soft, sales to cloud computing vendors are up 9%.[31]

Cloud computing disrupts the commercial software industry as well, for two reasons. First, many organizations aren't loyal to their software vendor; they're stuck. Cloud providers generally offer a variety of software options in each category, including open source and low cost "private label" software under their own brand. Opting to migrate to the cloud often puts an organization's software choices in play, encouraging switching.

The second source of disruption is elastic computing, and the notion that customers should only pay for what they use. The conventional software licensing model requires the customer to purchase a perpetual license or, at minimum, an annual term license; the cost of the license is sunk whether the customer uses it fully or not. The revenue model for many software vendors requires them to "stuff the channel" by loading up customers with software they will only partially use. If customers pay for what they use—and only what they use—they will pay a lot less.

Many commercial software vendors now support their software running in one of the top three cloud services. However, few offer elastic pricing, precisely because they fear that it will cannibalize their primary revenue model. In other words, commercial software vendors know that they have overlicensed their customers.

With cloud computing, it is now possible to build a complete business analytics platform entirely from services offered by the top three cloud computing vendors. With some technical skill, an analytics team can supplement vendor managed services with open source software to create a platform customized to meet the needs of any project. Moreover, this platform can scale out as needed to meet demand.

In Chapter Ten, we'll discuss such a platform in more detail, together with information about value-added managed services providers who provide complete business analytics "stacks" in the cloud.

---

[29]http://www.investors.com/news/technology/intel-struggles-to-separate-from-pc-q1-sales-miss-but-eps-tops/
[30]http://www.forbes.com/sites/greatspeculations/2016/04/21/emc-earnings-weakness-in-hardware-remains-non-core-businesses-drive-growth/#3a70214d74d4
[31]http://www.wsj.com/articles/mobile-shift-slams-techs-old-guard-1461195819

Machine Learning

Software That Learns

In Chapter Two, we surveyed the history of business analytics as a whole, noting that statistics and machine learning developed separately from data warehousing and business intelligence. In this chapter, we pick up where Chapter Two left off with a review of recent trends in machine learning.

Most of the key innovations in machine learning are distributed as open source software, which we discussed in Chapter Three. The discussion of scale-out architecture for machine learning extends the treatment of analytics in Hadoop covered in Chapter Four.

In Chapter Five, we covered Apache Spark, a distributed in-memory platform that is central to a discussion of distributed machine learning. We covered streaming machine learning briefly in Chapter Six and cloud-based machine learning in Chapter Seven.

Due to the significance of deep learning, we include a section covering this technology. We close the chapter with a survey of leading tools for modern machine learning.

© Thomas W. Dinsmore 2016
T.W. Dinsmore, *Disruptive Analytics*, DOI 10.1007/978-1-4842-1311-7_8

Recent Trends in Machine Learning

The most important trends affecting machine learning today are:

- Convergence of statistics and machine learning

- Growth of formal machine learning competitions

- Increased adoption of ensemble learning

- Development of scalable techniques for machine learning with Big Data

- Emergence of deep learning

Predicting the impact of these trends into the future requires some speculation; without question, though, they are affecting the machine learning discipline today.

Convergence

In 2001, Leo Breiman, professor emeritus of statistics at the University of California, Berkeley, wrote[1] of "two cultures" in predictive analytics. One culture, which he labeled as "data modelers," approached the predictive modeling problem by testing the hypothesis that the data conformed to one of several established functional forms.

The second culture, which he dubbed "algorithmic," approached the problem without assumptions and used machine learning tools to discover the model with the highest predictive power for the data at hand. Breiman used his own terminology, but it is clear that the "data modeling" label applied to the statistics community, and the "algorithmic" label to the machine learning community.

The "cultural divide" was even worse than Breiman described. Within the machine learning community there were numerous subcultures that developed around different core technologies, such as decision trees, neural networks, support vector machines, memory-based reasoning, and so forth.

Machine learning technologies developed separately from one another, with roots in different disciplines. Each developed its own language and tools. Practitioners developed skills and expertise in a single method, then vigorously argued that "their" method was better than all other methods. Each method had its own software implementation, which made comparison difficult.

---

[1]https://projecteuclid.org/euclid.ss/1009213726

Today, the debates are largely over and the cultural divide Breiman described is gone. For the most part, the "algorithmic" camp won; credentialed statisticians and actuaries freely use machine learning tools together with statistical techniques. Popular techniques, such as regularization, can't be easily assigned to one camp or another.

Regularization is a technique in machine learning to control overfitting, or the tendency of an algorithm to "learn" the characteristics of training data. Overfitting produces a model that predicts well on the training data but not on new data. Regularization controls for this problem by penalizing the loss function for each additional variable added to the model.

Three are three reasons for this convergence. First, the machine learning approach aligns with business needs better than the statistical approach. Breiman's "data modeling" culture defined success by methodological "correctness" and measured success with statistical "goodness of fit" measures. But most business leaders are not trained in statistics and have no interest in measures such as F-tests, T-tests, and R-squared; on the other hand, they immediately grasp measures such as accuracy and precision and understand testing predictions on historical data.

Second, the machine learning community has developed methods and procedures that control for concerns about bias or overfitting. Methods like out-of-sample and out-of-time testing, cross-validation, and partial dependency analysis are so powerful that they are used today with statistical techniques as well as with machine learning techniques.

Finally, data mining workbenches, introduced in the 1990s, combined different machine learning techniques with statistical techniques. These consolidated platforms made it easy for practitioners to test many different techniques and to choose the one best suited for the problem at hand.

Competition

Competitive machine learning, where teams and individuals compete to build the best model for prize money, has contributed greatly to the discipline. Competitions serve as laboratories for best practices in machine learning, and increase visibility of new techniques.

Since 1997, the Association for Computing Machinery's Special Interest Group on Knowledge Discovery and Data Mining (SIGKDD) has sponsored an annual competition called the KDD Cup. Each annual competition invites participants to complete a specific challenge, such as categorizing Internet search queries (2005); detecting breast cancer (2008); or predicting ratings on educational funding proposals (2014).

The competitions grow increasingly complex over time, from a straightforward classification problem in 1997 to the 2016 competition, in which teams compete to measure the relative influence of research institutions in a social graph. Increasingly, the challenges require entrants to blend multiple tools and techniques into an integrated solution.

The Netflix Prize was a highly visible contest that ran from 2006 to 2009. Netflix, the online DVD rental and video streaming service, offered $1,000,000 to the team that could beat Netflix' existing collaborative filtering algorithm by at least 10%. Netflix offered annual progress awards to the best performing team for the duration of the contest. For the contest, Netflix provided data sets for model training and for model evaluation, and specified the root mean squared error (RMSE) as the measure of model accuracy.

Netflix launched the competition on October 2, 2006. Within six days, a team beat Netflix's baseline.[2] At the end of the first and second competition years, Netflix awarded progress prizes, as no team had yet exceeded the 10% threshold.

Finally, in June 2009, two teams beat the 10% threshold. Over the course of the contest, 5,169 teams submitted 44,014 entries; the top two teams were closely matched, scoring RMSEs of 0.8554 and 0.8553, respectively. Netflix award the $1,000,000 to a team of seven researchers from Austria and the United States.

Inspired by the impact of the Netflix Prize, Anthony Goldbloom and Ben Hamner founded Kaggle in 2010 as a platform for predictive modeling and analytics competitions. Under Kaggle's model, a host organization sponsors a competition, defines the rules, and offers a prize. Kaggle provides a platform to host the data, accept submissions, maintains a leaderboard, and enforces the rules.

To date, Kaggle has hosted more than 200 public competitions for diverse sponsors, including Allstate, Caterpillar, GE, Heritage Health, Home Depot, Liberty Mutual, Merck, Prudential, Santander, and State Farm. First prizes range from knowledge, kudos, swag, and job opportunities to $500,000.

With more than a half-million registered users, Kaggle claims to have the world's largest community of data scientists. Kaggle tracks the performance of registered users on a leaderboard. In the absence of well-defined credentials for data scientists, the Kaggle leaderboard defines an elite community of experts.

Many other competitions support advances in specialized areas, such as handwriting recognition, traffic sign recognition, brain image classification, breast cancer diagnosis, and so forth. Successful efforts in these competitions contributed greatly to renewed interest in deep learning, which we discuss later in this chapter.

---

[2]http://www.hackingnetflix.com/2006/10/netflix_prize_r.html

It is difficult to overstate the impact of machine learning competitions. Competitions draw a great deal of interest from the machine learning community, and successful techniques are quickly disseminated. Moreover, the competitive environment demonstrates the value of collaboration and teamwork in advanced analytics and validates the crowdsourcing approach.

Ensemble Learning

As some researchers developed fundamentally new ways to train models, others found ways to improve models by combining techniques in various ways. Ensemble learning techniques use multiple models to produce an aggregate model whose predictive power is better than individual models used alone. These techniques are computationally intensive; growth in available computing power made ensemble learning accessible for mainstream users.

The many ways to combine models boil down to three: boosting, bagging, and blending. Boosting operates iteratively, successively building models on the errors of each previous model. ADABoost (Adaptive Boosting), introduced in 1995, is one of the most popular methods for ensemble learning. The ADABoost meta-algorithm operates iteratively, leveraging information about incorrectly classified cases to develop a strong aggregate model. With each pass, ADABoost tests possible classification rules and reweights them according to their ability to add to the overall predictive power of the model.

Leo Breiman developed a bagging algorithm in 1996. Bagging selects multiple subsamples from an original training data set, builds a model for each subsample, then builds a solution through averaging (for regression) or through a voting procedure (for classification). The principal advantage of bagging is its ability to build more stable models; its main disadvantage is its computational complexity and requirement for larger data sets. The growth of high-performance computing mitigates these disadvantages.

Stanford statistician Jerome H. Friedman introduced Gradient Boosting and a variant, Stochastic Gradient Boosting, in 1999. Gradient works in a manner similar to ADABoost, but uses a different measure to determine the cost of errors. Stochastic Gradient Boosting combines Gradient Boosting with random subsampling. In addition to improving model accuracy, this enhancement enables the analyst to predict model performance outside of the training sample.

In 2001, Breiman and Adele Cutler[3] proposed a technique they trademarked as "Random Forests". The Random Forests algorithm combines bagging (random selection of subsets from the training data) with the random selection of features, or predictors. The algorithm trains a large number of decision trees from randomly selected sub-samples of the training data set, then outputs the

---

[3] http://link.springer.com/article/10.1023%2FA%3A1010933404324

class that is the mode of the classes output by individual trees. The principal advantage of Random Forests compared to other ensemble techniques is that its models generalize well outside of the training sample. Moreover, Random Forests produces variable importance measures that are useful for feature selection.

Blended or stacked models are relatively new compared to the other techniques, but they have been used with great success in some highly visible competitions. A blended model leverages predictions from other models to develop an averaged prediction; the blended model outperforms any of the individual models. Blended models are more complex to train, since the analyst must train a number of based models first before building the blended model; they also take more time to produce predictions and may not be suitable for real-time applications.

Scaling to Big Data

A few software vendors developed software for statistics in the 1970s, 1980s, and 1990s. SAS Institute, through its strong partnership with IBM, established a reputation as the "enterprise" vendor for statistics through its commitment to the IBM mainframe. SPSS, spun off from the National Opinion Research Center at the University of Chicago in 1975, took a different approach, embracing the PC when it was introduced in 1984. SPSS delivered the first Windows-based statistical software in 1992 and grew rapidly by targeting the business user.

In the 1990s, SAS developed software that ran single-threaded on single machines. As analytic data sets grew larger in the 1990s and early 2000s, SAS hardware partners recommended larger and larger servers with more computing power to handle the expanded workload. Computing professionals call this approach "scaling up"—for more computing power, implement the software on a bigger computer.

Scaling up poses a number of issues as data sets grow larger. First, even the largest servers are too small for some projects. The limits of a single server forces analysts working on larger jobs to break the data into pieces and process it serially; as a result, large jobs can run for days—or even weeks.

The cost of the "big boxes" promoted by hardware vendors to enable scaling up is another issue. Large machines can run into the millions of dollars. Moreover, a computing architecture based on large machines is difficult to size and manage, because each increment to computing power is expensive. There is a tendency for "big box" architectures to behave like freeways: fast and expansive when new, but crowded and congested shortly thereafter.

Accordingly, most organizations have shifted toward a "scale-out" computing model, where applications run on many low-cost commodity servers. The scale-out model is easier to align with demand, because the computing infrastructure expands in small increments. Scale-out architecture is one of the primary reasons organizations adopt[4] Hadoop.

Some analytic tasks are easy to implement in a scale-out environment; we call these tasks *embarrassingly parallel* (see following note). Most model training algorithms are not embarrassingly parallel. Some are iterative, requiring multiple passes through the data; for others, item-level computations depend, in part, on other item-level communications and require interaction among distributed computing nodes.

An operation is *embarrassingly parallel* if computations on each data item are independent of computations on all other data items, and the product is a linear combination of distributed computations. Examples include SQL SELECT; scoring a linear model; and computing a statistical mean.

Tasks that are not embarrassingly parallel must be rewritten to run in a scale-out environment. This is expensive to do, and as we will show later in this chapter, there are just a few distributed engines on the market today.

Scaling to Big Data means working with larger data sets, but it also means working with diverse types (variety) and data in motion (velocity). We address machine learning with images, audio, video, speech, and other types of data under deep learning later, and we discussed streaming analytics in Chapter Six.

We tend to think of data volume in terms of items or rows in a table—a billion rows is a very big data set. However, the *width* of the data set—the number of columns, variables, or features—has a much greater impact on machine learning. Scientists have long recognized the *Curse of Dimensionality,* the computational problems associated with analyzing data with a large number of dimensions.

Columns, variables, features, and dimensions are closely related concepts that many people use interchangeably. A *column* is a set of values in a relational database table; a *variable* in computer programming is a symbolic name for a value that can change; a *feature* is a measurable property of an observed phenomenon; a *dimension* is a mathematical property. A thing has *features*; in a relational database, features map to columns; in a computer program, columns map to variables; in a mathematical discussion of the problem, variables map to dimensions.

---

[4]http://www.infoworld.com/article/2984534/application-development/hadoop-in-trouble-only-in-gartner-land.html

High-dimension data poses several problems for the analyst. The computational complexity of a problem increases rapidly with the number of dimensions. Additional dimensions also increase the number of possible ways a model can be specified, mandating more experiments to train and tune the model. Also, in the case of linear regression, a large number of dimensions increases the odds that some of them are correlated, leading to biased parameter estimates.

Machine learning researchers have developed several different approaches to *feature selection*, a pre-processing step implemented prior to model training. Stepwise regression, a method that iteratively adds or drops variables and re-trains the model, was a popular technique in the 1990s. However, it has fallen out of fashion[5] in favor of embedded methods, such as *regularization*, which progressively penalize additional variables.

Deep Learning

Three factors contributed to the growth of modern deep learning. The first of these is the introduction of general purpose computing on graphics processing units in the early 2000s. Graphics Processing Units (GPUs) are special chips originally developed to support computer gaming and image processing. GPUs are much more powerful than standard CPUs for certain types of tasks and have a highly parallel architecture. Support for floating point arithmetic and the development of APIs such as CUDA for general purpose computing make it practical to offload computing from CPUs to GPUs.

The Compute Unified Device Architecture, CUDA, created by GPU chip vendor Nvidia, is a software layer enabling programmers to use high-performance GPU chips for general-purpose computing.

A second factor contributing to the growth of deep learning was the development of knowledge and heuristics enabling practitioners to train the models effectively. Machine learning disciplines do not suddenly emerge by magic; successful application is the end result of a long process of experimentation and learning. Researchers struggled for years to solve the "exclusive-or" problem; due to the sheer complexity of deep learning models, it took years for the machine learning community to develop the skills and knowledge necessary to put the method to work.

The third factor is the huge expansion of digitized content—text, documents, images, audio, and video documented in the first chapter of this book. This huge expansion of digital content—what we now call Big Data—created entirely new applications for machine learning in areas such as sentiment

---

[5]http://www.lexjansen.com/pnwsug/2008/DavidCassell-StoppingStepwise.pdf

analysis, natural language processing, topic modeling, image recognition, image search, and speech recognition. Existing machine learning methods were less suited to these problems, which entail searching for hidden or latent patterns in massively "wide" sets of unlabeled data.

Deep learning reached an early milestone in 2007, when Geoff Hinton of the University of Toronto published[6] a seminal paper that outlined how a Deep Neural Network with multiple hidden layers could be trained layer by layer, thus breaking down the computational challenge into smaller and more tractable problems. Prior to that, research in speech and handwriting recognition had turned to so-called generative models.

Generative models are a class of statistical models that model relationships by learning the joint probability distribution for each data point; in contrast to discriminant models, which learn the conditional probability distribution. Examples of generative models include Gaussian Mixture Models, Hidden Markov Models, Latent Dirichlet Allocation, and Restricted Boltzmann Machines.

With expanded computing power and better methods, researchers working with deep learning started to show results. Beginning In 2009, Microsoft Research invested heavily in the application of deep learning to speech recognition and were able to significantly reduce[7] error rates compared to other methods.[8]

Similar efforts at Google in the field of image recognition also paid off. In 2012, *The New York Times* reported[9] that a Google Brain team used deep learning deployed over 16,000 computers to recognize unlabeled images among the millions of images in YouTube.

Thus, while interest[10] in neural networks has declined over the past decade, interest in deep learning has increased—markedly so since 2012 (see Figure 8-1). In that year, mainstream publications like *The New York Times*[11] and *The New Yorker*[12] wrote stories about how companies like Apple, Microsoft, and Google use deep learning to solve problems in speech recognition, image recognition, 3D object recognition, and natural language processing.

---

[6]http://www.cs.toronto.edu/~fritz/absps/tics.pdf
[7]http://ieeexplore.ieee.org/xpl/articleDetails.jsp?arnumber=6296526
[8]http://blogs.technet.com/b/inside_microsoft_research/archive/2015/12/03/deng-receives-prestigious-ieee-technical-achievement-award.aspx
[9]http://www.nytimes.com/2012/06/26/technology/in-a-big-network-of-computers-evidence-of-machine-learning.html
[10]Measured on Google Trends.
[11]http://www.nytimes.com/2012/11/24/science/scientists-see-advances-in-deep-learning-a-part-of-artificial-intelligence.html
[12]http://www.newyorker.com/news/news-desk/is-deep-learning-a-revolution-in-artificial-intelligence

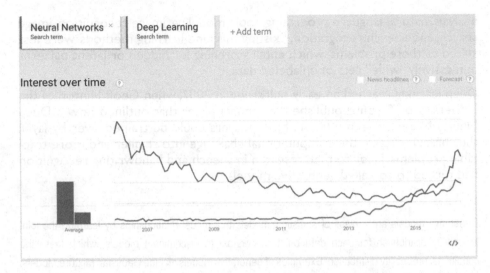

Figure 8-1. Search interest for neural networks and deep learning (Source: Google Trends)

In 2015, Google, Facebook, and Microsoft released open source deep learning frameworks to open source. We cover these frameworks later in this chapter.

Deep Learning Basics

In this section, we introduce the reader to some of the most important concepts in neural networks and deep learning.

Neural Networks

Since deep learning rests on the technology of neural networks, we begin with an introduction to key concepts in that field. This overview is necessarily simplified; there are volumes written on narrow subtopics in the field, and development is ongoing.

Animal brains are neural networks: networks of smaller cells, or neurons, linked together with synapses. As biologists studied animal brains, they built analog models of neural networks: physical devices that simulated brain function as well as possible using wires and light bulbs. These contraptions were artificial neural networks.

Neural networks as we know them today are symbolic representations of brain function coded in computer languages. A neural network represents a problem as a network of nodes ("neurons") connected by directed graphs ("synapses"). Like animal brains, they are able to "learn" and "remember". Figure 8-2 shows an example of a neural network.

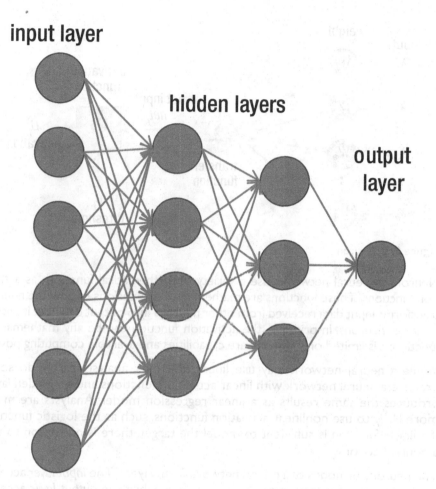

Figure 8-2. Neural network

Neuroscientists developed neural networks as a way to simulate animal learning. However, the methods they developed are broadly applicable in other fields.

In a neural network, each neuron accepts mathematical input, processes the inputs with a *transfer function*, and produces mathematical output with an *activation function*. Neurons operate independently on their local data and on input from other neurons. Figure 8-3 shows a neuron and its functions.

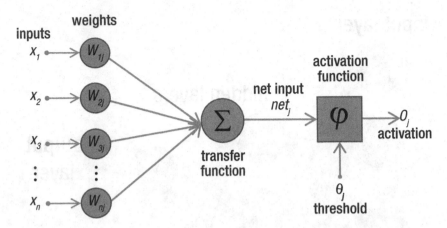

Figure 8-3. Neuron

Neurons in neural networks use a variety of mathematical functions as activation functions. These functions are mathematical expressions of how the neuron transforms input data received from other neurons into output data that it passes to other neurons. In principle, the activation function can be any mathematical function; it is limited only by software capabilities and available computing power.

While a neural network may use linear functions, analysts rarely do so in practice; a neural network with linear activation functions and no hidden layer produces the same results as a linear regression model. Analysts are much more likely to use nonlinear activation functions, such as the logistic function; if a linear function is sufficient to model the target, there is no reason to use a neural network.

The neurons or nodes of a neural network form layers. The *input layer* accepts mathematical input from outside the network, while the *output layer* accepts mathematical input from other neurons and transfers the results outside the network. A neural network may also have one or more *hidden layers* that process intermediate computations between the input layer and output layer. Deep neural networks are neural networks with at least two hidden layers. Figure 8-4 shows a deep neural network.

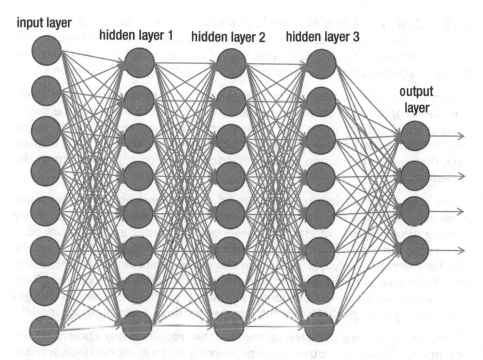

Figure 8-4. Deep neural network

The input and output layers of a neural network usually represent real-world facts: the input layer represents a vector of data we want to use as predictors, and the output layer represents a target variable.[13] Hidden layers, on the other hand, represent abstract concepts similar to factors in statistical factor analysis, except that they are not directly interpretable and simply serve to improve the accuracy of the model. Hidden layers enable neural networks to learn arbitrarily complex functions.

Practitioners classify neural network architectures according to the network topology, information flows within the network, mathematical functions, and training methods. The two most widely used architectures are:

Multilayer Perceptron. The Multilayer Perceptron (MLP) is a *feedforward* network; this means that neurons in one layer accept input from neurons in previous layers, but do not accept input from neurons in the same layer or subsequent layers. In an MLP, the parameters of the model include the weights assigned to each connection and to the activation functions in each neuron. Practitioners use a technique called *backpropagation* to train the network.

---

[13]http://neuralnetworksanddeeplearning.com/chap1.html.

Radial Basis Function Network. A Radial Basis Function (RBF) network uses radial basis functions, a particular type of mathematical function, as activation functions in the neurons. This type of neural network is well suited to function approximation, classification, and for modeling dynamic systems.

Analysts train a neural network by using one of many optimization algorithms. The backpropagation technique uses a data set in which values of the target (output layer) are known to infer parameter values that minimize errors. The method proceeds iteratively; first computing the target value with training data, then using information about prediction errors to adjust weights in the network.

There are several backpropagation algorithms; *gradient descent* and *stochastic gradient descent* are the most widely used. Gradient descent uses arbitrary starting values for the model parameters and computes an error surface; it then seeks out a point on the error surface that minimizes prediction errors. Gradient descent evaluates all cases in the training data set each time it iterates. Stochastic gradient descent works with a random sample of cases from the training data set. Consequently, stochastic gradient descent converges more quickly than gradient descent, but may produce a less accurate model.

Neural networks are complex techniques that require many choices by the practitioner. Relative to other machine learning techniques, however, artificial neural networks have four key advantages: an ability to automatically detect and model complex interactions among features; the ability to learn low-level features from minimally processed raw data; the ability to work with a large number of classes; and the ability to work with unlabeled data.

Taken together, these four strengths mean that artificial neural networks can produce useful results where other methods fail, and they have the potential to build more accurate models than other methods.

Unlabeled data lacks information about what it represents. A bit-mapped untagged photo, for example, is a stream of data characterizing the value of points in two-dimensional space, but does not include data about the subject of the picture.

Deep Learning Architectures

Building on neural networks, deep learning practitioners use complex new architectures that are well-suited to the key problems in content analytics posed by the tsunami of Big Data. Some of the most popular architectures include:

A **Restricted Boltzmann Machine (RBM)** is a shallow network with two layers: an input, or visible layer, and a hidden layer. Neurons or nodes in the input layer map to the features in the input data; for example, if a group of texts have 5,000 unique words, there will be one node in the input layer for each word. Nodes in the hidden layer represent relationships among the entities represented in the input layer; they are conceptually similar to factors in statistical factor analysis.

Deep-Belief Networks (DBN) are stacks of Restricted Boltzmann Machines. The hidden layer of each RBM serves as the input layer to the RBM above it in the stack. Analysts use DBNs to mine word vectors in text analytics, for image and video recognition and for voice recognition.

A **Deep Autoencoder** includes two symmetrical Deep-Belief Networks, each with four to five Restricted Boltzmann Machines arranged in layers. Deep autoencoders are useful for topic modeling, where the goal is to model abstract topics distributed across many documents. They are also used for data compression, and for image search applications, where images are first compressed into fixed-length numerical vectors.

Recursive Neural Tensor Networks have a tree structure with a neural network at each node of the tree. They are useful in text analytics, where they operate with word vectors.

Stacked Denoising Autoencoders (SDA) are stacks of another type of neural network called an *autoencoder*. The purpose of an autoencoder is to learn a representation of a set of data that reduces its dimensionality; an SDA has multiple hidden layers, each of which is an autoencoder. SDAs are useful for supervised document classification; for example, if we want to classify each document in a batch of documents into one of several groups for subsequent routing.

Convolutional Neural Networks (CNN) are a type of deep neural network inspired by the structure of the visual cortex in animals; they perform object recognition with images. Unlike multilayer perceptrons, whose neurons are fully connected, neurons in a CNN are locally connected to neurons in the immediate region. In image recognition, for example, one neuron represents one pixel in an image; in a CNN, that pixel may be connected to surrounding pixels, but not to a pixel in the far corner of an image. This approach is efficient when working with images.

Recurrent Networks (RNN) recognize patterns in sequences of data: time series data, handwriting, text, speech, or in genomes. Feedforward networks learn from data one case at a time, adjusting weights to minimize errors as they proceed through the data. RNNs, on the other hand, learn from both the current case and from the state of their own output as of the previous case, which serves as a kind of memory. Unlike a feedforward network, in an

RNN neurons may be connected to any other neuron in any layer, not just to neurons in previous layers.

Machine Learning in Action

This section presents 12 examples of machine learning in action.

Baidu.[14] In 2014, Baidu, a Chinese search engine company, announced development of a speech recognition system it calls Deep Speech.[15] Baidu claims that in noisy environments like restaurants, the system achieves an accuracy rate of 81%, a significant improvement over commercially available speech recognition software. The speech recognition system uses a Recurrent Neural Network (RNN). Baidu reported[16] using a computer cluster that is able to support deep learning models with about 100 billion neural connections (or synapses, not neurons).

Carolinas Healthcare System.[17] For hospitals, patient readmission is a serious matter, and not simply out of concern for the patient's health and welfare; Medicare and private insurers penalize hospitals with a high readmission rate, so hospitals have a financial stake in making sure that they only discharge patients who are well enough to stay healthy. The Carolinas Healthcare System (CHS) uses machine learning to construct risk scores for patients, which case managers use to make discharge decisions. This system enables better utilization of nurses and case managers, prioritizing patients according to risk and complexity of the case. As a result, CHS has lowered its readmission rate from 21% to 14%.

Cisco.[18] Marketers use "propensity to buy" models as a tool to determine the best sales and marketing prospects and the best products to offer. With a vast array of products to offer, from routers to cable TV boxes, Cisco's marketing analytics team trains 60,000 models and scores 160 million prospects in a matter of hours. By experimenting with a range of techniques from decision trees to gradient boosted machines, the team has greatly improved the accuracy of the models—that translates to more sales, fewer wasted sales calls, and satisfied sales reps.

---

[14]https://gigaom.com/2014/12/18/baidu-claims-deep-learning-breakthrough-with-deep-speech/
[15]http://arxiv.org/pdf/1412.5567.pdf
[16]http://www.bloomberg.com/news/articles/2014-09-04/baidu-builds-largest-computer-brain-for-online-queries
[17]http://www.healthcareitnews.com/news/predictive-analytics-lowers-readmissions
[18]http://www.datanami.com/2015/01/12/inside-ciscos-machine-learning-model-factory/

Comcast.[19] For customers of its XI interactive TV service, Comcast provides personalized real-time recommendations for content based on each customer's prior viewing habits. Working with billions of history records, Comcast uses machine learning techniques to develop a unique taste profile for each customer, then groups customers with common tastes into clusters. For each cluster of customers, Comcast tracks and displays the most popular content in real time, so customers can see what content is trending now. The net result: better recommendations, higher utilization, and more satisfied customers.

Dstillery.[20] Ad tech company Dstillery uses machine learning to help companies like Verizon and Williams-Sonoma target digital display advertising on real-time bidding (RTB) platforms. Using data collected about an individual's browsing history, visits, clicks, and purchases, Dstillery runs predictions thousands of times per second, handling hundreds of campaigns at a time; this enables the company to significantly outperform human marketers targeting ads for optimal impact per dollar spent.

GenomeDx Biosciences.[21] GenomeDx Biosciences is a startup in the business of genomic testing. To evaluate the efficacy of a genomic test in improving the diagnosis of prostate cancer, GenomeDx worked with major hospitals and medical schools to develop a clinical trial with 1,537 patients. The genetic test produced a vector of 46,000 features, far too many to analyze with conventional methods. Using a Deep Neural Network, GenomeDx built a classifier that predicted post-surgery outcomes for cancer patients more effectively than any other available method.

Jaguar Land Rover.[22] New cars built by Jaguar Land Rover have 60 onboard computers that produce 1.5 gigabytes of data every day across more than 20,000 metrics. Engineers at the company use machine learning to distill the data and to understand how customers actually use the vehicle. By working with actual usage data, designers can predict part failure and potential safety issues; this helps them to engineer vehicles appropriately for expected conditions.

---

[19]https://spark-summit.org/east-2015/talk/real-time-recommendations-using-spark

[20]http://bostinno.streetwise.co/channels/machine-learning-marketing/

[21]http://www.slideshare.net/0xdata/h2o-world-h2o-for-genomics-with-hussam-aldeen-ashab

[22]http://diginomica.com/2015/09/11/using-hadoop-inside-jaguar-land-rover-zurich-insurance-and-the-home-office/#.Vg1iqxNVhBc

Microsoft.[23] In March 2015, a Microsoft team published[24] a paper documenting results from their computer vision system, which is based on deep convolutional networks (CNNs). The team tested the system on the ImageNet 2012 classification data set, which contains 1.2 million training images, 50,000 validation images, and 100,000 test images. The task assigned to the system is to assign each image into one of 1,000 classes. The Microsoft system achieved a 4.94% error rate, which actually outperformed humans, who classified the images with a 5.1% error rate.

NBC Universal.[25] NBC Universal stores hundreds of terabytes of media files for international cable TV distribution; efficient management of this online resource is necessary to support distribution to international clients. The company uses machine learning to predict future demand for each item based on a combination of measures. Based on these predictions, the company moves media with low predicted demand to low-cost offline storage. The predictions from machine learning are far more effective than arbitrary rules based on single measures, such as file age. As a result, NBC Universal reduces its overall storage costs while maintaining client satisfaction.

PayPal.[26] Online payments company PayPal handles more than $10 billion in money transactions every month. At that volume, small improvements in fraud detection and prevention translate to significant bottom-line impact: each 1% improvement in prediction accuracy adds $1 million on operating contribution. Working with a data set of 160 million records and 1,500 features, the company's machine learning team continuously updates its fraud detection models, seeking small improvements. The company reports a "major leap forward" in its abilities since it started using nonlinear methods several years ago, and additional improvements since it started using deep learning three years ago. PayPal's deep learning algorithms can analyze thousands of latent features, such as time signals, actors, and geographic location, and have produced a 10% improvement over the previous champion fraud detection model.

Spotify.[27] A team at Spotify used a hybrid deep convolutional network to learn similarities and differences among songs based on spectrograms of the audio signal. Trained on 30-second tracks extracted from the million most popular songs on Spotify, the network learned to predict the latent representations of the songs obtained from a collaborative filtering model. (By doing so, Spotify can recommend playlists with little or no prior usage data.)

---

[23]http://blogs.technet.com/b/inside_microsoft_research/archive/2015/02/10/microsoft-researchers-algorithm-sets-imagenet-challenge-milestone.aspx

[24]http://arxiv.org/abs/1502.01852

[25]https://spark-summit.org/2015/events/use-of-spark-mllib-for-predicting-the-offlining-of-digital-media/

[26]https://gigaom.com/2015/03/06/how-paypal-uses-deep-learning-and-detective-work-to-fight-fraud/

[27]http://benanne.github.io/2014/08/05/spotify-cnns.html

U.S. Department of Energy.[28] Working with the Berkeley Lab, the National Energy Research Scientific Computing Center (NERSC) uses deep learning to analyze petabytes of data produced by climate simulation models. Using deep learning for pattern recognition, NERSC reports 95% accuracy detecting extreme weather events, such as tropical cyclones, atmospheric rivers, and weather fronts.

The New Machine Learning Software

In this section, we briefly describe scalable machine learning and deep learning software platforms, in four groups:

- Open source distributed engines
- Commercial distributed engines
- In-database libraries
- Deep learning frameworks

We do not include open source R and Python libraries, which we discussed in Chapter Four. While these languages are valuable developmental tools, they are not inherently scalable; R and Python users working with large data sets are best served by using one of the scalable engines listed here.

Many popular end user tools push processing down to one of the engines listed in this chapter. We cover those tools in Chapter Nine.

The distributed machine learning engines described in this section were originally developed to run either on clustered servers or special-purpose appliances; they were not originally designed to run in Hadoop. Under Hadoop 2.0 (after the release of YARN), all were quickly adapted to run in Hadoop under YARN.

Open Source Distributed Engines

There are just two open source general-purpose distributed engines for machine learning: Apache Spark and H2O. There are many other open source machine learning packages, such as Weka or Vowpal Wabbit, that do not support distributed model training. There are also machine learning tools, such as XGBoost, that support single algorithms. We limit the discussion to software that supports multiple algorithms.

---

[28]http://www.nersc.gov/news-publications/nersc-news/science-news/2015/nersc-berkeley-lab-explore-frontiers-of-deep-learning-for-science/

Apache Spark Machine Learning

Apache Spark MLlib is a machine learning library that runs on top of Spark. MLlib has two primary APIs; the original API that works with Spark RDDs (see Chapter Five) and a newer API that works with Spark DataFrames. (The PySpark and SparkR APIs also support machine learning functions.) While the Spark team continues to support the RDD-based API, all new development takes place in the DataFrames API.

The RDD-based API includes basic statistical tools, including summary statistics, correlations, stratified sampling, hypothesis testing, streaming significance testing, and random data generation. For machine learning, the library includes:

- Feature extraction and transformation, including tools for feature vectorization, text mining, standardization, normalization, feature selection, and vector transformation.

- Binary classification with linear support vector machines, logistic regression, decision trees, Random Forests, Gradient-Boosted Trees, and naïve Bayes classifier.

- Regression with linear least squares, Lasso, ridge regression, decision trees, Random Forests, Gradient-Boosted Trees, and isotonic regression.

- Dimensionality reduction, with Singular Value Decomposition (SVD) and Principal Component Analysis (PCA).

- Clustering with k-means, Gaussian Mixture, Power Iteration Clustering (PIC), Latent Dirichlet Allocation (LDA), bisecting k-means, and streaming k-means.

- Frequent pattern mining with FP-Growth, Association Rules, and PrefixSpan for sequence analysis.

- Collaborative filtering with Alternating Least Squares.

The API also includes a full set of statistics for model evaluation, including common metrics such as precision; recall, F-measure, ROC, and AUC. Users can also export PMML models for selected algorithms.

For developers who want to create their own algorithms, Spark exposes optimization primitives, including gradient descent, stochastic gradient descent, and the limited-memory Boyden-Fletcher-Goldfarb-Shanno (L-BFGS) algorithm.

The DataFrames-based API models a complete machine learning pipeline. Users work with DataFrames defined in Spark SQL rather than directly with RDDs. The basic elements in the API are *transformers* and *estimators*. Transformers are algorithms that perform operations one DataFrame to produce another DataFrame; for example, an operation that standardizes all variables in a data set. Estimators are algorithms that operate on a DataFrame to create a transformer; for example, a linear regression algorithm produces a linear model, which is a transformer that a user can apply to another DataFrame.

The library includes three types of prebuilt transformers:

- Feature extractors, which create features from raw data using algorithms like TF-IDF and Word2Vec.

- Feature transformers, which scale, convert, or modify features.

- Feature selectors, which select a subset from a larger set of features.

Machine learning functionality is rapidly expanding:

- Binary classification with logistic regression, decision trees, Random Forests, Gradient-Boosted trees, and Multilayer Perceptron.

- Multiclass classification through a "one versus all" algorithm used with binary classifiers.

- Regression with linear regression, decision trees, Random Forests, Gradient-Boosted trees, and survival regression.

- Clustering algorithms with k-means and Latent Dirichlet Allocation (LDA).

Spark Packages further extend Spark's machine learning with unique and innovative capabilities contributed by third parties. As of early 2016, there are more than 50 packages for machine learning.

For R users, the SparkR interface offers a selection of machine learning algorithms, including a Gaussian GLM model and Binomial GLM model. The Spark team has designed SparkR to operate in a manner similar to other R packages.

In addition to Spark's native machine learning libraries and Spark Packages, we note Apache SystemML, a module that runs on top of MapReduce and Spark.

SystemML is a declarative machine learning system developed by IBM and donated to the Apache Foundation; it is now an Apache Incubator project. Interacting with the software through Python and R APIs, users specify machine learning algorithms to run; SystemML generates optimized runtime plans for execution locally or in MapReduce or Spark.

As of early 2016, SystemML supports:

- Descriptive statistics, including univariate, bivariate, and stratified bivariate statistics.

- Classification techniques, including multinomial logistic regression, support vector machines, naïve Bayes, decision trees, and Random Forests.

- k-means clustering.

- Regression techniques, including linear regression, stepwise linear regression, generalized linear models, and stepwise generalized linear models.

- Matrix factorization techniques, including principal components analysis.

- Survival analysis, including Kaplan-Meier and Cox Proportional Hazards methods.

Users interact with SystemML through a high-level language (DML) with syntax similar to R or Python. DML includes linear algebra primitives, statistical functions, and ML-specific concepts. The algorithms, which are fully customizable, are dynamically compiled and optimized based on data and cluster characteristics.

Apache SystemML has a steadily growing code base and active contributor community.[29]

H2O

H2O is an open source distributed in-memory computing platform designed for deployment in Hadoop, in free-standing clusters, or in the cloud. H2O has its own distributed computing engine; it works with data in HDFS, S3, SQL, and NoSQL datastores, and with Apache Spark through the Sparkling Water interface.

Current functionality includes deep learning, generalized linear models, gradient boosted classification and regression, k-means clustering, naive Bayes classifier, principal components analysis, and Random Forests.[30] The software also includes tooling for data transformation, model assessment, and scoring. H2O exports scoring objects as Plain Old Java Objects (POJOs).

Users interact with the software through Java, Scala, Python, and R APIs, or through an easy-to-use web interface.

---

[29]https://www.openhub.net/p/apache-systemml
[30]https://thomaswdinsmore.com/2015/02/17/software-for-high-performance-advanced-analytics/

H2O.ai provides commercial support for the open source software. In July, 2014, H2O.ai received $8.9 million in Series A funding from a group of investors; subsequently, in November 2015, the company announced[31] a $20 million Series B round of funding. The company claims a number of public reference customers, including AT&T, Comcast, Kaiser Permanente, Progressive Insurance, Transamerica, Walgreens, and Zurich Insurance. There is a rapidly growing user community for H2O; H2O.ai claims more than 40,000 users in more than 5,000 organizations.

H2O has a large and steadily growing code base.[32]

Commercial Distributed Engines

Three software vendors offer commercially licensed distributed machine learning engines; SAS offers three different engines.

SAS High Performance Analytics

Industry leader SAS introduced SAS High Performance Analytics (HPA), in late 2012. HPA is a distributed in-memory analytics platform designed to run on specially built appliances from Oracle, Pivotal, or Teradata, and subsequently repurposed for clusters of commodity hardware. HPA serves as a back-end component for SAS Enterprise Miner and other SAS clients, enabling selected SAS procedures to run in distributed mode on clustered hardware.

While the product supports multiple databases, it lacks an open API and can only be called from SAS. HPA reads data into memory quickly through a parallel load, but does not keep data in memory and does not support high concurrency.

SAS introduced LASR Analytics Server in 2013 to serve as the back-end for a new visualization product (SAS Visual Analytics). LASR Analytics Server, unlike HPA, keeps data in memory and supports high concurrency. Neither architecture offers capabilities equivalent to a true in-memory database, such as durability guarantees or the ability to update data without reloading the entire data set.

In April 2016, SAS announced a third modern architecture, branded as SAS Viya, which the company positions as "open, elastic and scalable." As of August 2016, the software is in limited preview for existing SAS customers only, with planned general availability later in the third quarter of 2016.

---

[31]http://techcrunch.com/2015/11/09/h2o-ai-raises-20m-for-its-open-source-machine-learning-platform/
[32]https://www.openhub.net/p/h2o_by_0xdata

Microsoft R Server

Microsoft R Server is a commercially licensed software bundle that includes Microsoft R Open, an enhanced R distribution; integration and connectivity tools; and ScaleR, a library of distributed algorithms for predictive analytics with an R interface. The software runs on Linux; it can be deployed in Cloudera, Hortonworks, or MapR Hadoop distributions, or in Teradata Database. Microsoft offers the software on Windows through R Services for SQL Server 2016.

Microsoft R Server works with data in text files, HDFS, relational databases, SAS data sets, and other common formats. Capabilities supported in ScaleR include tools for data transformation, descriptive statistics, linear and logistic regression, generalized linear models, decision trees, ensemble models, and k-means clustering.[33] The software supports native model scoring and model export through PMML. The deployment interface supports integration with Tableau, Qlik, and custom web applications.

Skytree

Skytree is a Silicon Valley-based startup that develops and markets commercial software for machine learning. Skytree's core software began as an academic machine learning project (FastLab at Georgia Tech); the developers launched the company as a commercial software vendor in January 2013. The software runs under YARN on Cloudera, Hortonworks, MapR, and Amazon EMR, and integrates with Apache Spark to create what the company calls the Unified Machine Learning Platform.

The software supports data visualization, feature engineering, and machine learning algorithms for classification, regression, clustering, inference, and dimension reduction. Skytree claims an automated model selection capability, trademarked as AutoModel, which it is attempting to patent.

Users interact with the software through the Skytree Command Line Interface (CLI), Java and Python APIs, or a browser-based GUI.

In-Database Libraries

In-database machine learning libraries work inside relational databases, generally through table functions. Users interact with the machine learning library through SQL, or through applications that can pass SQL to the database through an open interface.

---

[33]https://thomaswdinsmore.com/2015/02/17/software-for-high-performance-advanced-analytics/

We include here only those machine learning libraries that can support multiple database platforms. This excludes from consideration the native machine learning tools built into IBM DB2, IBM Netezza, Microsoft SQL Server, Oracle Database, and Teradata Database. While the machine learning capabilities of these databases may be useful (especially in organizations that are fully committed to the database platforms), most organizations are better off investing in capabilities that are not tied to single vendors.

Apache MADlib

Apache MADlib is an open source library of machine learning algorithms designed to operate in massively parallel databases, without data movement. Development started in 2010 as a collaboration between researchers at UC Berkeley and data scientists at EMC Greenplum (now Pivotal Software).

Pivotal donated the software assets to the Apache Software Foundation in 2015, and the project entered Apache incubator status. While the project seeks to broaden its contributor base, most recent commits come from two Pivotal employees.

The project explicitly supports PostgreSQL, Pivotal Greenplum Database, and Pivotal Hawq; in principle, users can implement the library in any database that supports UDAs, such as Impala.[34].

The MADlib algorithms operate as table functions in databases; users invoke them through SQL. MADlib also supports feature extraction from text and low-rank matrix factorization together with a number of utilities for discovery, validation, and model implementation. Machine learning capabilities include 10 different regression methods, linear systems, matrix factorization, tree-based methods, association rules, clustering, topic modeling, text analysis, time series analysis, and dimension reduction techniques.

Commercial support for MADlib is unclear at this time. Most MADlib users are customers of Pivotal Software, and that company provided consulting and technical support. Dell recently acquired EMC, Pivotal Software's parent company; meanwhile, shifting project governance from Pivotal to Apache will likely expand the user and contributor base.

---

[34]https://www.facebook.com/notes/facebook-engineering/scaling-apache-giraph-to-a-trillion-edges/1015

Fuzzy Logix DB Lytix

DB Lytix, a commercial software offering from Fuzzy Logix, is library of more than 800 functions for machine learning and advanced analytics. Functions run as database table functions in relational databases (Informix, MySQL, Netezza, ParAccel, SQL Server, Sybase IQ, Teradata Aster, and Teradata Database) and in Hadoop through Hive. DB Lytix also runs in GPU devices; Fuzzy Logix offers the Tanay Zx appliance for GPU-based analytics.

Users invoke DB Lytix functions from SQL, R, through BI tools, or from custom web interfaces. Functions support a broad range of machine learning capabilities, including feature engineering, model training with a rich mix of supported algorithms, plus simulation and Monte Carlo analysis. All functions support native in-database scoring. The software is highly extensible, and Fuzzy Logix offers a team of well-qualified consultants and developers for custom applications.

In November, 2015, Fuzzy Logix announced[35] that it raised $5.5 million in venture capital.

Deep Learning Frameworks

A recent article in VentureBeat lists[36] no fewer than 15 different software frameworks for deep learning. We describe some of the most promising open source projects in the following sections.

CNTK

The Computational Network Toolkit (CNTK) is a product of Microsoft Research. Microsoft developed CNTK to improve computer speech recognition[37], and it uses[38] it in products such as Windows Cortana, Skype Translator and Project Oxford Speech APIs. Microsoft released[39] the software to open source in January 2016.

---

[35]http://www.businesswire.com/news/home/20151103006068/en/Fuzzy-Logix-Raises-5.5-Million-Science-Ventures

[36]http://venturebeat.com/2015/11/14/deep-learning-frameworks/

[37]http://news.microsoft.com/features/speak-hear-talk-the-long-quest-for-technology-that-understands-speech-as-well-as-a-human/

[38]http://blogs.technet.com/b/inside_microsoft_research/archive/2015/12/07/microsoft-computational-network-toolkit-offers-most-efficient-distributed-deep-learning-computational-performance.aspx

[39]http://blogs.microsoft.com/next/2016/01/25/microsoft-releases-cntk-its-open-source-deep-learning-toolkit-on-github/

Like TensorFlow and Theano, CNTK represents networks as a graph that represents mathematical operations as nodes; the edges between nodes represent multidimensional data arrays. This approach allows users to invent new network architectures and layer types. The software runs on standard CPUs as well as graphical processing units (GPUs), machines with multiple GPUs, and distributed on a cluster of multi-GPU machines.

Users interact with CNTK by first creating a configuration file, then running the software from a command-line interface. There is no API.

CNTK is based on C++, so developers can compile trained models and deploy them across platforms.

TensorFlow

TensorFlow is the second generation of a machine learning system developed by Google scientists and engineers. Google uses its first generation system, called DistBelief in a number of Google applications, including search, voice search, photo recognition and video matching[40]. DistBelief learns concepts such as "cat" from unlabeled YouTube images, and improves speech recognition in the Google app. It won[41] ImageNet's Large Scale Visual Recognition Challenge in 2014.

Google engineers simplified and rebuilt the DistBelief code to create TensorFlow; in November, 2015, Google released[42] a reference implementation of TensorFlow to open source under an Apache license. Although Google's internal version of TensorFlow can distribute workload over clustered machines, the open source version runs on a single machine only[43]. It supports GPUs through CUDA extensions. Supported operating systems include Linux and Mac OS. The package supports Python and C++ APIs.

TensorFlow models machine learning operations in the form of a graph that represents mathematical operations as nodes; the edges between nodes represent multidimensional data arrays ("tensors" in Google terminology). Although Google engineers developed TensorFlow for deep learning, the system can be generalized to other machine learning operations.

---

[40]http://techcrunch.com/2015/11/09/google-open-sources-the-machine-learning-tech-behind-google-photos-search-smart-reply-and-more/
[41]http://googleresearch.blogspot.com/2014/09/building-deeper-understanding-of-images.html
[42]http://www.wired.com/2015/11/google-open-sources-its-artificial-intelligence-engine/
[43]http://www.wired.com/2015/11/googles-open-source-ai-tensorflow-signals-fast-changing-hardware-world/

Developers can compile trained models and deploy them on a variety of devices. However, as of Spring 2016, they cannot be deployed on Windows.

In February, 2016, Google released[44] TensorFlow Serving, a software package designed to simplify the deployment of trained models. The software works natively with TensorFlow, and it can also support other tools.

Theano

Theano is a Python library for numerical computation developed by scientists at the Université de Montréal. It allows users to efficiently define, optimize, and evaluate mathematical expressions with multi-dimensional arrays. While Theano's capabilities are not limited to deep learning, its transparent support for GPU processing makes it a popular platform for deep learning practitioners.

Like CNTK and TensorFlow, Theano represents neural networks as a symbolic graph. Theano was the first to do so, and due to its maturity most state of the art network architectures are available on the platform.

Since Theano lacks a low-level interface, it is less suitable for production applications due to Python overhead. In performance benchmarks, Theano lags[45] other deep learning frameworks, such as TensorFlow, Caffe, and CNTK. It supports single GPU machines only.

DL4J

Deeplearning4j (DL4J) is an open source computing framework written in Java that supports deep learning algorithms. Skymind, a small San Francisco based startup, leads software development for the project and provides commercial support. Skymind distributes Dl4J under an Apache 2.0 license.

DL4J is a distributed and multi-threaded framework; it is integrated with Hadoop and Spark and trains models within the cluster. In a distributed environment, DL4J shards, or splits a large data sets and passes the shards to worker nodes for execution. Each node trains a model on its local data; DL4J then iteratively averages the parameters to produce a single model.

Written in Java, DL4J also offers APIs to the related Scala and Clojure languages. It supports standard CPUs and GPUs.

---

[44]http://venturebeat.com/2016/02/16/google-open-sources-tensorflow-serving-for-deploying-machine-learning-models/
[45]http://blogs.microsoft.com/next/2016/01/25/microsoft-releases-cntk-its-open-source-deep-learning-toolkit-on-github/

Caffe

Caffe is a deep learning framework developed by the Berkeley Vision and Learning Center (BVLC) and released under an open source BSD license. Stemming from BVLC's work in vision and image recognition, Caffe's core strength is its ability to model a Convolutional Neural Network (CNN).

Caffe is written in C++. Users interact with Caffe through pycaffe, a Python package, or through a command-line interface. Deep learning models trained in Caffe can be compiled for operation on most devices, including Windows.

In February 2016, Yahoo released[46] CaffeOnSpark, a package that enables Spark users to embed Caffe deep learning into Spark processes. Yahoo has successfully applied CaffeOnSpark to image recognition problems, significantly improving image recognition accuracy by training with millions of photos from the Yahoo Webscope Flickr Creative Commons data set.

Apache SINGA

Currently in Apache Incubator status, Apache SINGA is an open source distributed deep learning platform for training deep learning models on large data sets. Researchers from the National University of Singapore lead the development team.

The platform currently supports feed-forward models, convolutional neural networks, restricted Boltzmann machines, and recurrent neural networks. The project includes a stochastic gradient descent algorithm for model training.

SINGA currently supports GPU processing on a single node. Training on a GPU cluster is under development.

Torch

Torch is an open source scientific computing framework developed by a team of engineers from Facebook, Google, and Twitter. First released in 2002, the software is available under a BSD license.

Torch supports packages for multi-dimensional tensors, neural networks, optimization, 2D and 3D plotting, file manipulation, and image processing. Users can build Convolutional Neural Networks (CNN), including temporal convolution, as well as Recurrent Neural Networks (RNN).

---

[46]http://yahoohadoop.tumblr.com/post/139916563586/caffeonspark-open-sourced-for-distributed-deep

The package offers an API in LuaJIT, an easy-to-use scripting language, so defining new network architectures is simple. While LuaJIT is easy to learn and use, it is more difficult to integrate into a production pipeline.

The New Machine Learning

The war of words between devotees of statistical techniques and machine learning techniques is largely over. Except for a few holdouts, the machine learning camp and its pragmatic approach has won the day. Practitioners freely mix techniques from the two camps using methods and procedures perfected by the machine learning community.

Highly visible competitions contribute to this convergence. Results from entries using different techniques are transparent to everyone. Teams using the most powerful techniques win; teams that are unwilling to part with traditional techniques lose.

Ensemble learning techniques, first developed in the 1990s, are increasingly mainstream. Several factors contribute to this development: high quality open source software implementations, increased availability of computing power at reduced cost, and visible successes in machine learning competitions.

As analysts grapple with Big Data, software developers have introduced distributed machine learning engines with a scale-out architecture. The transition to distributed engines is a discontinuity in the market. Single-threaded server-based machine learning software is increasingly seen as obsolete, creating opportunities for startups. Building distributed computing platforms is expensive and difficult, so many developers leverage existing open source frameworks such as Apache Spark.

Deep learning has emerged as a practical technique to address the most challenging problems in machine learning, such as speech and image recognition. Declining costs of computing and the emergence of high-performance low-cost GPU-based platforms have accelerated the adoption and use of deep learning.

There is growing interest in self-service machine learning for business users. We cover this in Chapter Nine, together with automated machine learning.

Self-Service Analytics

Hype and Reality

About 15 years ago, the author attended a sales presentation by a vendor touting an automated predictive analytics tool. His value proposition was, "buy our software and you can fire your SAS programmers".

Unfortunately for that sales rep, every customer in the room was a SAS programmer.

The story underscores a basic problem for those who believe that software can democratize analytics: the people who care the most about analytics, and are most passionate about it, are not afraid to learn analytic programming languages like SAS, Python, and R.

Moreover, because they are accountable for the veracity and validity of what they deliver, they *demand* tools that give them control over the entire process. That is why experts on the analytics job market insist that coding skills are absolutely necessary to do the job.[1]

So much for the Citizen Data Scientist idea.

---

[1]http://www.kdnuggets.com/2014/11/9-must-have-skills-data-scientist.html

© Thomas W. Dinsmore 2016

T. W. Dinsmore, *Disruptive Analytics*, DOI 10.1007/978-1-4842-1311-7_9

Commercial vendors have touted their analytics software as "easy to use" or "self-service" for three decades, and yet, adoption for tools other than Microsoft Excel remains low. There are a number of reasons for this:

- Many managers have minimal felt need or interest in analytics.

- For managers who do value analytics, it's relatively easy to delegate the hands-on work.

- Making software easy to use is one thing; making data easy to access and navigate is an entirely different matter. After two decades of data warehousing, enterprise data remains messy, incomplete, or irrelevant to managers' questions.

Meanwhile interest in analytic programming languages is booming, and the job market for data scientists is robust. That is because serious problems in analysis require serious people to perform them—people who are motivated to dig deeply into data.

In this chapter, we review the logic of self-service analytics—where it makes sense, and where it doesn't. We describe the distinctly different user personas in organizations and discuss the role of experts in advanced analytics. We close the chapter with a survey of six innovations in self-service analytics.

The Logic of Self-Service

Vendor-driven discussions of self-service analytics often inhabit a magical world where everyone has the same skills and interests. Managers are best served by a realistic view of the actual user personas in their organization, and the use cases where self-service analytics make sense.

The User Pyramid

In most organizations, people vary widely in their analytic skills—from those with little or no skill on one end of the spectrum, to world-class experts on the other. Many things cause these differences: background, education, training, organization role, and intrinsic motivation.

In large organizations, analytics users tend to form a "pyramid," as shown in Figure 9-1. Those with the lowest purely analytic skills, whom we label as "consumers," tend to be the largest group by far; in large organizations, there can be thousands of them. We note that this is purely analytic skills; people can have highly advanced skills in other areas, but limited training or interest in the analysis of data.

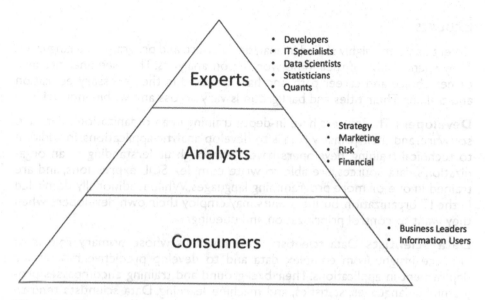

Figure 9-1. The user pyramid

Analytics experts, on the other hand, tend to be few in number. However, they drive disproportionate value through analytics.

Commercial software vendors tend to target casual users with "easy to use" applications. Casual users are the largest audience, offer the potential to sell more "seats," and are less loyal to existing tools. Power and expert users, on the other hand, tend to be very loyal to their existing tools. They have invested years to develop skills, have mastered their chosen tool, and are not attracted to "easy to use" tools.

Roles in the Value Chain

Serious discussions about analytics in the organization should begin with the recognition that users have diverse needs and are not all the same. This may seem obvious, but one frequently hears vendors speak of their tools as complete solutions for the enterprise, as if all users have the same needs.

A *user persona* is a model that describes how a *class* of user interacts with a system. Of course, every user is different and so is every organization. There are startups in Silicon Valley where "business users" work actively with SQL and Python; there are also companies where "business users" struggle with Microsoft Excel. We present this model not to stereotype people, but as a framework for managers to understand the needs of their own organizations.

Experts

Expert users are highly skilled in analytic software and programming languages. They spend 100% of their time working on analytics. They see analytics as a career choice and career path, and have invested in the necessary education and training. Their titles and backgrounds vary across and within industries.

Developers. These users have in-depth training in an organization's data and software, and their primary role is to develop analytic applications. In addition to technical training, developers have a thorough understanding of an organization's data sources, are able to write complex SQL expressions, and are trained in one or more programming languages. While traditionally domiciled in the IT organization, business units may employ their own developers when they want to control prioritization and queuing.

Data Scientists. Data scientists are individuals whose primary role is to produce insight from complex data and to develop predictive models for deployment in applications. Their background and training encompasses programming languages, statistics, and machine learning. Data scientists tend to come from an engineering or computer science background, and they prefer to work with programming languages such as Scala, Java, or Python. They tend to have a strong preference for open source analytics; the best data scientists actively contribute to one or more open source projects and may participate in data science competitions.

Analytic Specialists. The analytic specialist holds a position such as statistician, biostatistician, actuary, or risk analyst and often holds a degree in an academic discipline with historical roots in advanced analytics. They understand statistics and machine learning, and they have considerable working experience in applied analytics. Analytic specialists prefer to work in a high-level analytic programming language such as SAS or R, which they prefer over software packages with GUI interfaces. Their work product may be a management report, charts and tables, or a predictive model specification.

Analysts

These users are highly skilled in the use of end user software for analytics, such as Excel, SPSS, or Tableau. They use analytics actively in their work, but they do not create production applications. They tend to identify with a business function, such as marketing or finance, and see analytics as a means toward that end.

Strategic Analysts. Strategic analysts' primary role is to perform ad hoc analysis for senior executives; they may be domiciled within a business unit or within a team dedicated to C-team support. Strategic analysts know their organization and industry well, and they are familiar with data sources. They are able to perform simple SQL queries and to use SQL together with other tools. They prefer tools with a graphical user interface. Strategic analysts' work product leans toward charts, visuals, and storytelling.

Functional Analysts. The primary role of functional analysts is an analytic job function, such as credit analyst or marketing analyst. These roles require some analytic skill as well as domain knowledge. Functional analysts prefer tools that are relatively easy to use, with a graphical interface. They may have some training in statistics and machine learning. Like strategic analysts, they may be able to perform simple SQL queries. They are skilled with Microsoft Office and prefer to work with tools that integrate directly with Office. Functional analysts' work product may be a spreadsheet, a report, or presentation.

Consumers

Information consumers have minimal tool-related skills and prefer information presented in a form that is easily retrieved and digested.

Business Leaders. Business leaders are keenly interested in the organization's performance metrics, which they require to be timely, accurate, and delivered to a mobile device or browser. They may be interested in some limited drill-through capabilities, but rarely want to spend a great deal of time searching for information.

Information Users. Information users are employees who need information to perform a specific job role, such as handling customer service calls, reviewing insurance claims, performing paralegal tasks, and so forth. Their role in a business process defines their needs for information. While the information user may not engage with mathematical computation, they are concerned with the overall utility, performance, and reliability of the systems they use.

The Role of Experts

Software vendors tout their products as "easy to use". This is not new. In the 1980s and 1990s, analytics vendor SPSS positioned its Windows-based interface as the easy alternative to SAS, which did not offer a comparable UI until 2004, when it introduced SAS Enterprise Guide. In the 1990s and early 2000s, Cognos claimed to target the business user in contrast to more complex products like Business Objects and MicroStrategy.

Vendor claims to the contrary, self-service analytics is an elusive goal. Large enterprises have thousands of users for their BI tools, but the vast majority are "consumers" who use the information contained in reports, views or dashboards developed by specialists. Most organizations still maintain specialist teams whose sole responsibility is developing reports or OLAP cubes for others to use.

There are two main reasons for the persistence of BI specialists:

Consistency. Measuring performance remains the leading use case for business intelligence. Most organizations want consistent measurement across functions and do not want teams to measure themselves, or "game" the metrics.

Data Integration. Few organizations have achieved the data warehousing "nirvana" envisioned by theorists. While BI tools have well-designed user-facing "interfaces," the "back-end" that integrates with data sources remains as messy and complicated as the sources themselves.

Expanding use of Hadoop has exacerbated the data integration problem for BI, at least temporarily. Most conventional BI platforms did not work with Hadoop 1.0. Startups like Datameer and Pentaho tried to fill this vacuum, with limited success; but specialized tools just for Hadoop are unsatisfactory for enterprises seeking to standardize on a single BI platform.

Hadoop poses a problem even for relatively skilled users. An analyst accustomed to working interactively in SQL on a data warehouse appliance will struggle to perform the same analysis in Hive or Pig on Hadoop. Hadoop's tooling for interactive queries has greatly improved in the last several years, as documented in Chapters Four and Five, companies that invested early in Hadoop added new layers of specialists with the necessary skills.

Predictive analytics also remains the domain of specialists even though easy-to-use tools have been available for years. This is especially so in strategic and "hard-money" applications, such as fraud detection and risk management, where the quality of a predictive model can mean the difference between business success and business failure.

Analytic experts provide executives with what auditors call the *attest function*[2] —an independent certification that the analysis is correct and complete. For predictive models, the expert attests that the model predicts well, minimizes false positives and false negatives appropriately, and does not encourage adverse selection. Few executives have the necessary training and knowledge to verify the quality of complex analysis by themselves. The need for independent attestation is a primary reason that organizations outsource strategic analysis projects to consultants.

In short, organizations do not employ experts and specialists for their skill with analytic programming languages. They employ them primarily for their domain expertise, for their ability to take ownership for the analysis they provide, and for their willingness to be held accountable for its validity.

---

[2]http://www.investopedia.com/terms/a/attest-function.asp

Drawing an analogy to medicine, robots are now able to perform the most complex heart surgery. This does not eliminate the need for cardiologists, although it may change the nature of the job. Patients are not likely to entirely entrust a diagnosis to a machine, nor do machines have the bedside manner that patients value.

A Balanced View of Self-Service

While performance measurement and strategic analytics will remain in the hands of experts and specialists, self-service analysis makes the most sense for two use cases: discovery and business planning.

Discovery and Insight. Before initiating action, managers want to understand the basic shape of a problem or opportunity. This naturally leads to questions such as:

- How many customers do we have in New Jersey?
- How many shoppers bought our brand of dog food last week?
- What is the sales trend for our stores in the Great Lakes region?

Discovery is ideally suited to self-service analysis for three reasons. First, handing the work to a specialist slows the process down. Specialists are often backlogged; a task may only take the specialist an hour to complete, but due to previous requests in the queue, the requestor waits a week for results.

Second, handing the analysis to a specialist creates potential misunderstandings. The requestor must prepare a specification; requests grow increasingly detailed, and specialists develop a habit of doing exactly what the requestor requests rather than developing insight into the problem the requestor is trying to solve.

Third, discovery is iterative by nature; the answer to one question prompts additional questions. The specialist model discourages iteration, since every cycle requires another request, another wait in the queue, and more potential for misunderstandings.

From a functional perspective, interactive queries and visualization are the key requirements for discovery tools. A flexible back-end, with easy integration to many different data sources, is absolutely necessary. Additional capabilities managers may require for discovery include simple time series analysis, simple predictive modeling, basic content analytics, and a mapping capability.

Business Planning. After a manager identifies an opportunity, the next step is action planning; this, in turn, raises many tactical questions. For example, once a decision is made to conduct a marketing campaign among current customers to stimulate purchase of a particular product, managers may want to know:

- How many customers purchase this product?

- What is the average purchase volume among these customers?

- How many customers purchased the product in the past twelve months but not the past three months?

As with discovery, managers assign a premium to speed and self-service as they develop business plans. In planning, however, managers need quantification and numerical analysis more than simple visualization. Interesting trends and hypotheses are less interesting than hard numbers at this point.

Predictive analytics play a greater role in business planning. The manager is concerned with forecasting the impact of a particular decision:

- What is the expected response and conversion rate?

- What credit losses can we expect?

- If we do nothing, what attrition rate do we expect?

Hence, self-service tools for predictive analytics can play a role in business planning, provided that they are fully transparent and "idiot-proofed". An "idiot-proof" tool has built-in constraints and guidelines that prevent a naive user from developing spurious insights.

Innovations in Self-Service Analytics

In the section that follows, we discuss six self-service innovations:

- **Data visualization:** Tableau combines a simple interface for basic visualization with a powerful data access engine.

- **Data blending:** Alteryx and ClearStory Data take very different approaches to the problem of blending data from diverse data sources.

- **BI on Hadoop:** AtScale and Platfora demonstrate two distinct ways to deliver BI on Hadoop.

- **Insight-as-a-Service:** Domo helps executives avoid the IT bottleneck.

- **Business user analytics**: KNIME, RapidMiner, and Alpine offer distinct approaches to scalable analytics for business users.

- **Automated machine learning**: DataRobot delivers an automated machine learning platform.

In previous chapters, we highlighted open source tools. All of the software we profile in this chapter is commercially licensed. In each of the six categories, we profile one or more vendors that exemplify the innovation.

Data Visualization

A picture is worth a thousand words.

Data visualization seeks to discover, communicate information, and persuade through statistical graphics, plots, and infographs. As a discovery tool, visualization helps the analyst quickly identify meaningful patterns in data that would be difficult to find through numerical analysis alone. As a communications tool, visualization makes complex relationships clear and comprehensible. Visualization is a powerful tool for persuasion.

Visualization is also a great way to lie and mislead people. Data visualization is no more "scientific" than its user intends it to be. Every manager should understand the rhetoric of visualization if only to identify deception.

Like most things in the world of analytics, visualization is hardly new. Statisticians have long understood the value of data visualization. In 1977, John Tukey, founding chair of the Statistics faculty at Princeton University, introduced the idea of the box plot. Box plots, shown in Figure 9-2, are a convenient way to compare multiple data distributions.

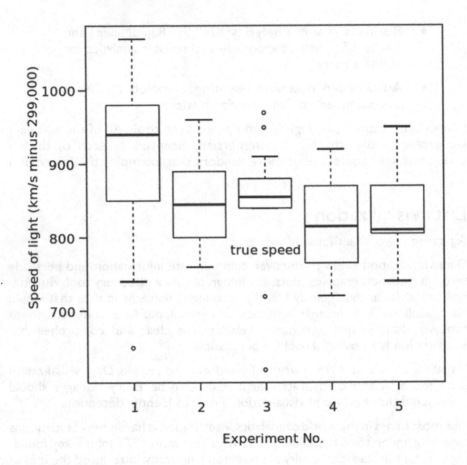

Figure 9-2. Box plot

Edward R. Tufte, professor of Political Science, Statistics, and Computer Science at Yale University, published *The Visual Display of Quantitative Information* in 1982. Tufte self-published the book due to lack of interest from publishers; today, it remains a best seller on Amazon.com. Tufte drew on examples of visualization dating back to 1686, noting that many of the best visuals pre-date the computer era, when artists drew graphs by hand.

Stephen Few, a data visualization consultant and author of *Information Dashboard Design,* argues[3] that there are just eight types of visual messages:

- Time series charts, with values showing change through time

- Ranking of values ordered by quantity

---

[3]http://www.perceptualedge.com/articles/misc/Graph_Selection_Matrix.pdf

- The relationship of parts to the whole
- The difference between two sets of values, such as fore-cast revenue and actual revenue
- Counts or frequencies of values by interval
- Comparison of two paired sets of values to show cor-relation (or the lack thereof)
- Nominal comparison of values for a set of unordered items
- Geospatial depiction of data, where values or measures are displayed on a map

In preliminary data analysis, statisticians and predictive modelers use scat-terplots and frequency distributions to understand relationships in the data. Scatterplots show linear and non-linear relationships between two variables, which can expedite model development. Simple graphics are also an excellent data quality check, as an analyst can instantly identify problems in the data from a few visuals.

Statisticians and market researchers use visualization to convey complex find-ings. Correlations among many variables can be hard to interpret when pre-sented as a table of numbers; presented as a heat map, as shown in Figure 9-3, patterns are easier to grasp.

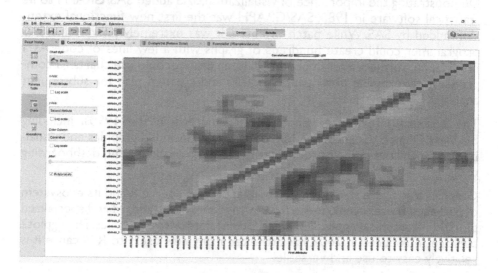

Figure 9-3. Correlation heat map created in RapidMiner

Techniques like decision trees are popular because they are easy to visualize. Figure 9-4, for example, shows the characteristics of people who survived the Titanic sinking: women in the top two classes, younger women in third class, and boys under 13 survived at a much higher rate than other passengers.

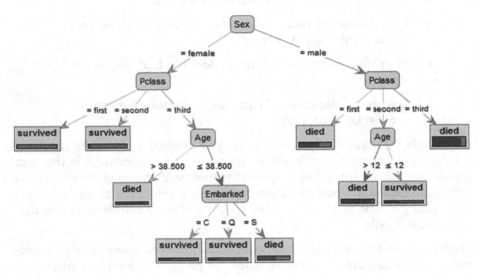

Figure 9-4. Titanic survivor decision tree (Source: RapidMiner)

Demonstrating the importance of visualization, SAS added SAS/GRAPH to its statistical software in 1980. SAS/GRAPH was the first new functional extension added by SAS. While batch oriented, it was extremely powerful; a SAS programmer could create hundreds of graphs with a few lines of code. As a first step in a project after importing data, an analyst could use charts to comprehensively examine potential predictors and create a plan for subsequent analysis.

The business intelligence vendors that emerged in the 1990s, including Business Objects, Cognos, and MicroStrategy, all included graphics and visualization capabilities in their products together with reporting and dashboarding tools.

Graphics also figure strongly in the appeal of open source R and its ecosystem of packages. R users can choose from a wide range of standard and specialized plots, build graphical applications, and publish interactive graphics. The ggplot2 package, introduced in 2007, has attracted many new users to R because it is relatively easy to use and versatile.

It is impossible to discuss the recent emergence of self-service visualization without mentioning Tableau. Researchers at Stanford University founded

Tableau Software in 2003 to commercialize an innovative visualization tool they had developed at the university. Tableau grew steadily, reaching revenues of $62 million in 2011; then, it took off, growing revenues tenfold to $654 million.

In so doing, Tableau passed Actuate, Hexagon, Panorama Software, Autodesk, TIBCO, Information Builders, ESRI, Qlik Tech, and MicroStrategy to take the sixth position on IDC's Business Intelligence and Analytics Tools Software ranking. Tableau also passed industry stalwarts FICO, Infor, Adobe, and Informatica to assume eighth place in IDC's overall business analytics software industry ranking.

Tableau went public in May 2013 at a valuation of $3 billion. From its founding to its IPO, Tableau created[4] more value for its investors than all but three other startups from 2009 through 2014.

Tableau's success is surprising when you consider that its graphical capabilities are no greater than most competing business intelligence tools, and considerably less than what a user can do in SAS and R. The key differences between Tableau and other tools are simplicity and data source connectivity. Conventional BI tools are designed to integrate with an organization's data warehouse. While their graphics capabilities are not difficult to use, they require complex configuration for each data source. This makes them inflexible, requiring a high level of skill for ad hoc analysis.

For tools like SAS and R, visualization is a two-step process: one step to retrieve data, the second to create visuals. While these tools are powerful and highly flexible, they require expertise to use successfully. Tableau's core innovation[5] is a query language called VizQL, or Visual Query Language. VizQL combines SQL and a language for rendering graphics, so that ad hoc visualization requires only a single step. Tableau combines its query language with connectivity to an extraordinarily large collection of data sources, including Microsoft Excel and Access; text files; statistical files from SAS, SPSS, and R; relational databases; NoSQL datastores; Hadoop; Apache Spark; enterprise applications, including SAP and Salesforce; Google Analytics; and many others.

Arguably Tableau is successful not because it does so much, but because it does a few things very well, and the things it does well are exactly what users need. Tableau's simplicity makes it easy to use. Combined with its powerful data source connections, Tableau works very well as an ad hoc discovery tool in diverse and complex data. Under conventional data warehousing theory, this use case should not exist, since data warehousing theory calls for the consolidation of data into a single datastore. Tableau's extraordinary business success demonstrates the degree to which conventional data warehousing theory no longer applies.

---

[4]https://www.cbinsights.com/blog/capital-efficient-tech-exits-top-25/
[5]http://dl.acm.org/citation.cfm?id=1142473.1142560

Data Blending

Data blending tools enable a business user to blend and cleanse data from multiple sources. They come with rich facilities to access disparate data sources, select data, transform the data, and combine it into a single dataset for analysis. Most have some capability to analyze the blended data as well.

According to data warehousing theory, there should be no need for end user data blending tools; in principle, data warehousing processes should perform all of the necessary processing steps, presenting the end user with data that is already cleansed, standardized, and in the form needed for analysis.

- In many organizations, budget constraints prevent the data warehousing team to keep up with the explosion of data.

- Even well-funded data warehousing teams have substantial backlogs leading to extended delays bringing new sources into the warehouse.

- Many analyses are ad hoc and do not warrant investment in permanent data warehousing feeds.

As an example of the last point, many marketing programs use external vendors, and the campaign may only run once or twice. A marketing analyst seeking to prepare an analysis of the campaign must merge data provided by the external vendor with data from the organization's data warehouse to prepare a complete report.

A number of startups offer data blending tools, including Alteryx and ClearStory Data.

In 1997, three entrepreneurs founded a consultancy branded as SRC LLC; the company offered custom solutions for mapping and demographic analysis. Two years later, SRC won a bid to be the technology provider for the U.S. Bureau of the Census; over the next several years, the company developed several new software products for geospatial analysis.

SRC launched Alteryx in 2006. Alteryx, a software package offering a unified environment for the analysis of spatial and non-spatial data simplified the task of blending data from multiple databases; streamlined spatial analysis; and enabled users to publish integrated reports with maps, charts, tables, and graphs.

In 2010, the SRC founders rebranded the company as Alteryx Inc. to focus exclusively on this product[6]. Alteryx has raised a total of $163 million in three rounds of venture capital; the most recent round of funding, for $85 million, closed in October 2015.

As of June, 2016, Alteryx Analytics is in Release 10. The Alteryx Designer environment enables a business user to build workflows to prepare, blend, and analyze data from a wide range of sources and data types, including

- Data warehouses and relational databases
- Cloud and enterprise applications
- Hadoop and NoSQL datastores
- Social media platforms
- Packaged data from suppliers like Experian, Dun & Bradstreet, and the U.S. Bureau of the Census
- Microsoft Office and statistical software packages

For deeper analysis, Alteryx offers basic descriptive and predictive analytics built in open source R, as well as geospatial analytics. Users can export analysis in Microsoft Office formats, Adobe PDFs, HTML, and other common formats. Alteryx interfaces with leading visualization tools, such as Tableau, Qlik, Microsoft Power BI, and Salesforce Wave.

Figure 9-5 shows a view of the Alteryx Designer desktop.

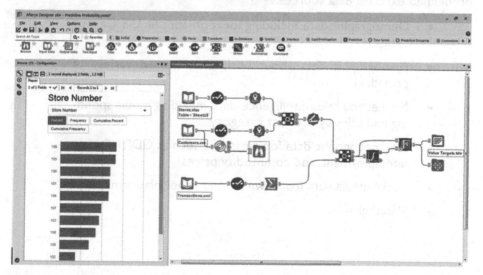

Figure 9-5. Alteryx Designer

---

[6]http://www.alteryx.com/press-releases/src-llc-is-now-alteryx

The Alteryx Server edition runs on Microsoft Windows Server. Alteryx Server supports scalable analytics through push-down SQL, which transfers user requests to the datastore for native execution without data movement. Release 10 supports push-down SQL for Amazon Redshift, Apache Hive, Apache Impala, Microsoft SQL Server, and Azure SQL Data Warehouse, Oracle Database, Apache Spark SQL, and Teradata Database.

ClearStory Data takes a very different approach to data blending. Founded in 2011, ClearStory Data released its ClearStory product to the market in 2013. ClearStory is an in-memory visualization and collaboration application combined with an inference engine and data blending capability. The platform runs exclusively on Apache Spark (discussed previously in this chapter and in Chapter Two).

Building on Spark's data-ingestion capabilities, ClearStory provides organizations with the ability to integrate disparate internal and external data sources. Supported internal sources include:

- Relational databases, including Oracle, SQL Server, Amazon Redshift, MySQL, and PostgreSQL

- Hadoop

- Files in a variety of formats

- APIs for enterprise applications, such as Salesforce

Through partnerships with data providers, ClearStory provides a number of predefined external data sources:

- Demographic data, including location-specific U.S. Census data

- Firmographic data about businesses from a variety of providers

- Market and sales intelligence data, including media spending and sales by product category

- Macroeconomic data for measures such as GDP, inflation, unemployment, and commodity prices

- Social media data from Twitter and other platforms

- Weather data

Once a data source is registered with ClearStory, the application's data infer-ence engine gathers key statistics profiling the shape of the data, as well as information about its structure and semantics. When a user requests analysis, ClearStory uses this information to recommend additional data based on the problem the user is trying to solve. ClearStory's data blending engine matches data with common dimensions, enabling the user to combine data from dis-parate sources.

Through late 2015, investors have provided ClearStory Data with $30 million in venture capital. The most recent[7] round, in March 2014, was a $21 million Series B funding led by DAG Ventures, with Andreessen Horowitz, Google Ventures, Khosla Ventures, and Kleiner Perkins Caufield & Byers participating.

BI on Hadoop

As noted in Chapter Four, organizations are investing heavily in Hadoop. Hadoop's significantly lower costs compared to traditional data warehouses make it an attractive alternative, especially for data whose value is not yet established.

However, Hadoop is much harder to use than traditional data warehouses. For end users accustomed to using business intelligence tools with a data warehouse, Hadoop is almost impossibly difficult to use. Even tools like Hive and Pig, which are easier to use than MapReduce, are only suitable for an advanced user.

As the volume of data residing in Hadoop expands, there is a growing need for business user tools that can work with the data. AtScale and Platfora are two startups with very different approaches to this problem. AtScale delivers a middle layer that enables existing business intelligence tools to work with Hadoop data. Platfora, on the other hand, creates a dedicated data mart to support its own end user tools. We discuss these two startups next.

Founded in 2013 by Yahoo veterans, AtScale emerged from stealth in April 2015; simultaneously, it announced a $7 million "A" round of funding. Unlike the other BI startups profiled in this chapter, AtScale does not offer its own BI end user client. Instead, AtScale operates on the principle that most organiza-tions already have BI tools in place, so it works in the background to make these tools work with a Hadoop datastore.

In theory, most BI tools can connect directly to Hive tables or Spark DataFrames through the JDBC API. In practice, unless the data is already structured and aggregated with all of the needed measures, dimensions and

---

[7]http://techcrunch.com/2014/03/31/clearstory-raises-21m-from-dag-ventures-kpcb-a16z-to-bring-data-intelligence-to-the-masses/

relationships, the user will have to switch back and forth from the BI tool to Hive, Pig, or a programming API. Few business users have the skills needed to do this, so the organization must assign a developer, move the data elsewhere, or implement and maintain a special-purpose "BI-on-Hadoop" tool.

An *edge node* is a server on the periphery of a Hadoop cluster, which is typically used to broker interactions with other applications. An edge cluster is similar to an edge node, but consists of a cluster of servers on the periphery of the Hadoop cluster rather than a single server.

The AtScale engine resides on an edge node in a Hadoop cluster. With user input through the web-based AtScale Cube Designer, AtScale interacts with the Hive metastore to create and maintain a virtual dimensional data model, or "cube". Users can specify hierarchies for drill-down, calculated fields and other dimensions as needed to represent the business problem at hand. The foundation data does not change or move, so users can specify different cubes based on the same data for different purposes. BI tools such as Microsoft Excel or Tableau submit SQL or MDX requests to AtScale through ODBC, JDBC, or OLE DB. End users work directly from tools like Excel, as shown in Figure 9-6.

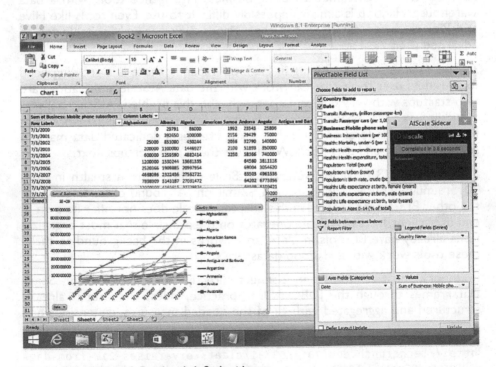

Figure 9-6. Microsoft Excel with AtScale sidecar

AtScale develops and submits an optimized query through an available SQL engine (such as Hive on Tez, Spark SQL, or Cloudera Impala) and returns the results to the BI tool for further processing and end user display. Users can elect whether to retain or drop any aggregate tables created. AtScale provides a facility for managing aggregate tables and a capability to schedule cube generation and updates.

AtScale supports popular BI tools: Tableau, Microsoft Excel, Qlik, Spotfire, MicroStrategy, PowerBI, JasperSoft, SAP Business Objects, and IBM Cognos. It works with the Cloudera, Hortonworks, MapR, and HDInsights Hadoop distributions; Hive on Tez, Spark SQL, and Cloudera Impala SQL engines; and a broad selection of data storage formats, including Parquet, RC, ORC, Sequence, text files, and Hive SerDe. For security, the software offers role-based access control to selectively grant access to data to users across departments and organizations. The application maintains an audit trail of queries executed, so the organization can track request volume, data requested, and run times.

Ben Werther, a veteran of Siebel, Microsoft, and Greenplum, founded Platfora in 2011 with $5.7 million in funding[8] from venture capitalists led by Andreesen Horowitz. The company emerged from stealth mode in October 2012 and closed[9] a $20 million "B" round shortly thereafter. The company raised[10] an additional $38 million in March 2014. (Author's note: on July 21, 2016, cloud-based software vendor Workday announced plans to acquire Platfora for an undisclosed amount.)

Platfora offers an end-to-end data warehousing and BI platform that runs on an edge cluster next to Hadoop. Platfora Server is a distributed in-memory engine; it operates on copies of the data extracted from Hadoop.

Before end users work with the data, an administrator defines structured Platfora datasets from source data. In addition to defining data structure, the administrator defines access permissions for the dataset.

End users, who work from a browser interface, work with the defined datasets to specify a view of the data they need. Platfora translates these requests into MapReduce or Spark jobs, submits them for execution, and writes the results back to HDFS in Platfora's proprietary file format. It also retains a copy of the view locally on disk and registers metadata about the view in a catalog.

Platfora presents a user-friendly interface that is accessible to a business user. As the user explores and analyzes the data, Platfora generates in-memory queries against the local copy of the view. If the data to be queried exceeds available memory in the Platfora cluster, the query spills to disk or fails.

---

[8]http://techcrunch.com/2011/09/08/andreessen-horowitz-leads-5-7m-round-in-analytics-platform-for-hadoop-data-platfora/
[9]http://www.finsmes.com/2012/11/platfora-closes-20m-series-funding.html
[10]http://techcrunch.com/2014/03/19/big-data-analytics-company-platfora-raises-38m/

For the most part, Platfora works only with data already loaded into Hadoop; it has a limited capability to pull small datasets from other sources. Platfora works with most Hadoop distributions; it can use data stored in HDFS, Hive, MapR FS, uploaded files and also with Amazon Web Services' S3 file system.

Insight-as-a-Service

Many executives are frustrated by what they perceive as a lack of responsiveness and poor service quality from their IT organization. Motivated by a need for speed, they seek out services that can immediately provide them with the performance metrics and insight they desire.

In the past, bypassing the IT organization was difficult because IT physically controlled all of the data. Today, with many business processes delivered through hosted services and Software-as-a-Service, a considerable amount of data already resides in the cloud. Moreover, as functional leaders increasingly control technology spend, they effectively "own" the data.

Vendors with pre-packaged cloud-based solutions address the needs of these executives. One such vendor, Domo, demonstrates the power of the insight-as-a-service concept.

Domo, a startup located in Salt Lake City, Utah, takes a radically different approach to BI. Instead of promoting another set of tools, Domo positions itself as a management solution for busy executives frustrated by the delays and limitations of conventional BI tools. Domo provides these executives with the means to completely bypass IT bottlenecks with a packaged cloud-based Software-as-a-Service delivery model.

Josh James, a successful entrepreneur, founded the company in 2010. James co-founded Omniture, a successful web analytics startup, in 1996, and led the company through a successful $1.8 billion sale of the company to Adobe Systems in 2009. In late 2010, James acquired a small company offering visualization software and renamed the combined entity Domo. With ample capital—$484 million in eight rounds from 45 investors, including Andreessen Horowitz, Fidelity Investments, Jeff Bezos, Morgan Stanley, and T. Rowe Price—Domo remained in stealth mode for almost five years, developing and improving its offering.

A company operating in stealth mode does not disclose information about its product or service to the public; it does this so potential competitors cannot anticipate its offering and to allow sufficient time to develop a marketable product. The company may disclose information to investors or consultants, but only under strictly enforced nondisclosure agreement. Since they do no marketing, companies operating in stealth mode rarely have revenue or customers, so a company may need substantial funding to remain in stealth mode for an extended period.

Domo emerged[11] from stealth mode in April 2015 with a highly developed product. Around a core of standard BI functions (including queries, reports, dashboards, and alerts), Domo offers pre-built role-based and industry-based solutions and apps designed for decision support. The user-facing capabilities operate on a modified MPP columnar database running on Amazon Web Services.[12]

Domo has also pre-built more than 350 connectors to data sources to expedite data integration. This library of connectors includes the most widely used databases and applications; for marketing alone, there are 51 connectors, including Adobe Analytics, Facebook, Google AdWords, HubSpot, IBM Digital Analytics, Klout, Marketo, Salesforce, SurveyMonkey, Twitter, Webtrends, YouTube, and many others.

Combined with a facility for secure data transfer from on-premises systems, these pre-built connectors and solutions enable Domo to promise rapid time to value. Moreover, Domo's focus on offering an integrated and customizable role-based decision-making solution differentiates it from conventional BI tools.

When it emerged from stealth in April 2015, Domo claimed to have more than 1,000 paying customers and annual sales of $50 million in 2014.

Business User Analytics

Commercial vendors compete actively to deliver software for predictive analytics that is both easy to use and powerful. This is not a new phenomenon; the following list shows five such products and the year each was introduced.

- Angoss KnowledgeSeeker (1984)
- SAS JMP (1989)
- Dell Statistica (1986)
- IBM SPSS Modeler (1994)
- SAP InfiniteInsight (1998)

Three relatively new products deserve more detailed discussion. KNIME and RapidMiner, introduced in 2006, and Alpine, introduced in 2011. KNIME and RapidMiner operate under an open core model; each offers an open source edition together with commercially licensed extensions. All three are suitable for Big Data, offering push-down integration with Hadoop; Alpine also offers push-down integration with selected data warehouse appliances.

---

[11]http://venturebeat.com/2015/04/08/domo-comes-out-of-stealth-after-five-years-and-raises-200m-at-2b-valuation/
[12]https://web-assets.domo.com/blog/wp-content/uploads/2015/08/Domo_-Impact-Report_-451-Research_-7-May-2015.pdf

KNIME (rhymes with "lime") is an open source platform for data integration, business intelligence, and advanced analytics. The platform, based on Eclipse and written in Java, features a graphical user interface with a workflow metaphor. Users build pipelines of tasks with drag-and-drop tools and run them interactively or in batch execution mode. Figure 9-7 shows a view of the KNIME Analytics Platform desktop.

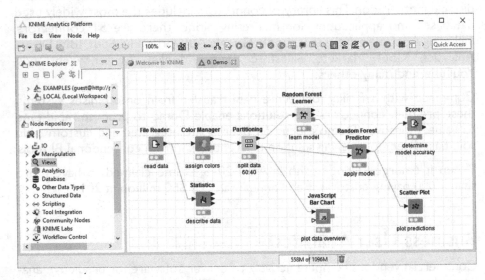

Figure 9-7. KNIME Analytics Platform desktop

KNIME.com AG, a commercial enterprise based in Zurich, Switzerland, distributes the KNIME Analytics Platform under a free and open source GPL license with an exception permitting third parties to use the API for proprietary extensions. The company is privately held and does not disclose details of its ownership. There is no record of venture capital investment in the company.

The free and open source KNIME Analytics Platform includes the following capabilities, all implemented through the graphical user interface:

- Data integration from text files, databases, and web services

- Data transformation

- Reporting through the bundled open source Business Intelligence and Reporting Tool (BIRT)

- Univariate and multivariate statistics

- Visualization using interactive linked graphs

- Machine learning and data mining

- Time series analysis
- Web analytics
- Content analytics, including text and image mining
- Graph analytics, including network and social network analysis
- Native scoring, as well as PMML export and import
- Open API for integration with other open source projects and with commercial tools

KNIME.com AG also distributes a number of commercially licensed extensions offering additional capabilities not included in the open source platform. They include:

- Enhanced tools for building workflows
- Authoring tools to create custom extensions
- Collaboration tools for file and workflow sharing
- Server-based tools for enhanced security, collaboration, scheduling, and web access
- Connectors enabling push-down execution in Apache Hive, Apache Impala, and Apache Spark
- Tools to manage job execution on clustered servers

The KNIME Analytics Platform operates in-memory on single machines running Linux, Windows, or Mac OS. The software is multi-threaded to use multiple cores on a single machine. Server and cloud extensions run on the same operating systems. KNIME.com AG supports the Hive and Impala extensions on Cloudera, Hortonworks, and MapR Hadoop distributions; the company supports the Spark extension on Cloudera and Hortonworks.

Since KNIME buffers data to disk, it can in theory handle arbitrarily large datasets that exceed memory. Disk buffering, however, affects performance and can lead to longer runtimes.

The KNIME Big Data Extensions enable KNIME users to push SQL workloads into Hadoop through Apache Hive or Apache Impala, and to run Apache Spark applications. The Spark Executor serves as an interface to the Spark MLlib package, enabling users to run classification, regression, clustering, collaborative filtering, and dimension reduction tasks in Spark. The software includes a PMML 4.2 interface for prediction in Spark, and also enables the user to perform data preprocessing and manipulation with Spark.

KNIME.com AG offers commercial technical support for the extension software. For the open source KNIME Analytics Platform, it offers extensive product documentation and a community forum for troubleshooting. The company also certifies partners and resellers who offer consulting and support services.

RapidMiner is a mixed commercial and open source software platform for advanced analytics developed and distributed by RapidMiner, Inc. of Cambridge, Massachusetts. Started as predictive analytics project at the Technical University of Dortmund, RapidMiner has expanded its capabilities to span the entire advanced analytics process, from data integration to deployment.

RapidMiner, Inc. launched in 2006 (under the corporate name of Rapid-I) to drive software development, support, and distribution. The company moved its headquarters to the United States in 2013 and rebranded as RapidMiner. Since then, it has secured $36 million in venture capital in three rounds. The most recent, a $16 million "C" round, closed[13] in January 2016.

Under a model it calls "business source," RapidMiner distributes three software editions:

- **Basic edition**: Available under a free and open source license.
- **Community edition**: Available under a free commercial license with registration.
- **Professional edition**: Commercially licensed under a paid subscription.

The core RapidMiner platform (Basic edition) includes:

- Data ingestion from Excel, CSV, and open source databases, data blending, and data cleansing functions.
- Diagnostic, predictive, and prescriptive modeling functions.
- R and Python script execution.

The free Community edition also includes:

- A small cloud instance.
- Community technical support.

---

[13]https://www.pehub.com/2016/01/rapidminer-raises-16-mln/

- Access to the RapidMiner marketplace.

- The "Wisdom of Crowds" feature. RapidMiner collects detailed usage information from its user community and leverages this information to provide recommended actions.[14]

The Professional edition also includes:

- Reusable building blocks and processes for the Design Studio.

- Access to commercial databases, cloud data sources, NoSQL datastores, and other file types.

- A larger cloud instance.

RapidMiner offers a workflow interface that enables the user to construct complex analytic "pipelines," as shown in Figure 9-8.

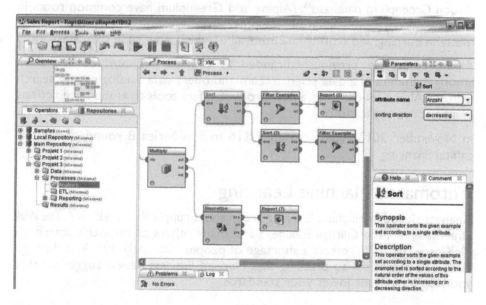

Figure 9-8. RapidMiner

In addition to the desktop version, RapidMiner commercially licenses software for servers and for Hadoop (branded as "Radoop"). The server version supports collaboration, performance, and deployment features; Radoop supports push-down integration with Hive, MapReduce, Mahout, Pig, and

[14]https://rapidminer.com/wisdom-crowds-guiding-light/

Spark. RapidMiner supports Radoop with Cloudera CDH, Hortonworks HDP, Apache Hadoop, MapR, Amazon EMR, and Datastax Enterprise.

RapidMiner has implemented about 1,500 functions in Spark, and it permits the user to embed SparkR, PySpark, Pig, and HiveQL scripts. RapidMiner supports the software with the open source Apache Hadoop distribution, plus distributions from Cloudera, Hortonworks, Amazon Web Services, and MapR; DataStax Enterprise NoSQL database; Apache Hive and Apache Impala; and Apache Spark.

RapidMiner's key strengths are its easy-to-use interface, broad functionality, and strong integration with Hadoop. While RapidMiner's predictive analytics and optimization features are strong, its visualization and reporting capabilities are limited, which makes it unsuitable for some users.

Alpine Data Labs, founded in 2011, offers Alpine ML, software with a visual workflow-oriented interface and push-down integration to relational databases, Hadoop, and Spark. Alpine claims support for all major Hadoop distributions and several MPP databases, though in practice most customers use Alpine with Pivotal Greenplum database[15]. (Alpine and Greenplum have common roots in the EMC ecosystem). Alpine ML supports data ingestion, feature engineering, machine learning, and scoring functions, all of which execute in the datastore.

Alpine Enterprise, previously branded as Chorus, facilitates collaboration among members of a data science team and offers data cataloging and search features. Alpine Touchpoints, a new product, offers tools to embed predictions in interactive applications.

In November 2013, Alpine closed a $16 million Series B round of venture capital financing.

Automated Machine Learning

Analysts skilled in machine learning are in short supply. VentureBeat[16], *The Wall Street Journal*[17], the *Chicago Tribune*,[18] and many others all note the scarcity; a McKinsey report[19] projects a shortage of people with analytical skills through 2018. The scarcity is so pressing that *Harvard Business Review* suggests[20] that you stop looking, or lower your standards.

---

[15]https://thomaswdinsmore.com/2015/02/17/software-for-high-performance-advanced-analytics/

[16]http://venturebeat.com/2015/03/17/why-data-scientists-and-marketing-technologists-are-the-hottest-jobs-of-2015/

[17]http://blogs.wsj.com/cio/2014/11/10/for-cios-universities-cant-train-data-scientists-fast-enough/

[18]http://www.chicagotribune.com/business/ct-indeed-survey-0514-biz-20150514-story.html

[19]http://www.mckinsey.com/features/big_data

[20]https://hbr.org/2014/09/stop-searching-for-that-elusive-data-scientist/

Can we automate the work data scientists do? In *IT Business Edge,* Loraine Lawson wonders[21] if artificial intelligence will replace the data scientist. In *Forbes,* technology thought leader Gil Press confidently asserts[22] that the data scientist will be replaced by tools; Scott Hendrickson, Chief Data Scientist at social media integrator Gnip, agrees.[23]

Data mining web site KDnuggets, which caters to data scientists, recently published[24] a poll of its own members that asked *when will most expert level data scientist tasks be automated?* Only 19% of respondents believe they will never be automated; 51% said they would be automated within the next 10 years.

Automated modeling techniques are not new. In 1995, Unica Software introduced Pattern Recognition Workbench (PRW), a software package that used automated test and learn to optimize model tuning for neural networks. Three years later, Unica partnered with Group 1 Software (now owned by Pitney Bowes) to market Model 1, a tool that automated model selection over four types of predictive models. Rebranded several times, the original PRW product remains as IBM PredictiveInsight, a set of wizards sold as part of IBM's Enterprise Marketing Management suite[25].

KXEN, a company founded in France in 1998, built its analytics engine around an automated model selection technique called structural risk minimization.[26] The original product had a rudimentary user interface, depending instead on API calls from partner applications; more recently, KXEN repositioned itself as an easy-to-use solution for marketing analytics, which it attempted to sell directly to C-level executives. This effort was modestly successful, leading to sale of the company in 2013 to SAP for an estimated[27] $40 million.

Early efforts at automation from Unica, MarketSwitch, and KXEN "solved" the problem by defining it narrowly; limiting the scope of the solution search to a few algorithms, they minimized the engineering effort at the expense of model quality and robustness. Second, by positioning their tools as a means to *eliminate* the need for expert analysts, they alienated the few people in customer organizations who understood the product well enough to serve as champions[28].

---

[21]http://www.itbusinessedge.com/blogs/integration/will-artificial-intelligence-replace-the-data-scientist.html

[22]http://www.forbes.com/sites/gilpress/2012/08/31/the-data-scientist-will-be-replaced-by-tools/

[23]https://blog.gnip.com/data-scientist-vs-data-tools/#

[24]http://www.kdnuggets.com/polls/2015/analytics-data-science-automation-future.html

[25]https://www.datarobot.com/blog/automated-machine-learning-short-history/

[26]http://www.svms.org/srm/

[27]https://451research.com/report-short?entityId=79713

[28]https://www.datarobot.com/blog/automated-machine-learning-short-history/

Data scientists say[29] they spend 50-80% of their time on data wrangling. In theory, this means organizations can mitigate the shortage of data scientists by improving data warehousing and management practices; in practice, this is not easy to do. Data warehousing is expensive, and data scientists often support forward-looking projects that move too fast for the typical data warehousing organization. Most data scientists see data wrangling as necessary and unavoidable, and to a considerable degree they are right.

Automation can, however, reduce the time, cost, and pain of data wrangling. Built-in integration with widely used data sources, for example, minimizes the time and cost to extract and move data. Interfaces to data warehousing and business intelligence platforms enable data scientists to directly leverage data that is already cleansed, minimizing duplicate effort. Features that automatically detect and handle missing data, outliers, complex categorical fields, or other "problematic" types of data enable data scientists to work with data "as is," and eliminate the need for manual processing.

Beyond basic data cleansing and consolidation, the requirements for data transformation ("feature engineering") depend entirely on the algorithm to be used for model training. Some algorithms, for example, will only work with categorical predictors, so any continuous variables in the input data set must be binned; other algorithms have the opposite requirement. Automated feature engineering must be linked to automated model specification and selection, since the two are intrinsically linked.

The best way to determine the right algorithm for a given problem and data set is a test-and-learn approach, where the data scientist tests a large number of techniques and chooses the one that works best on fresh data. (The No Free Lunch[30] theorem formalizes this concept.) There are hundreds of potential algorithms; a recent benchmark study tested[31] 179 for classification alone.

When computing power was scarce and expensive, modelers dealt with this constraint by limiting the search to a single algorithm—or a few, at most. They defended this practice by minimizing the importance of predictive accuracy or by defending the use of one technique above all others. This led to endless unempirical flame wars between advocates of one algorithm or another.

Cheap and pervasive computing power ends these arguments once and for all; it is now possible to test the power of many algorithms, selecting the one that works best for a given problem. In high-stakes hard-money analytics—such as trading algorithms, catastrophic risk analysis, and fraud detection—small improvements in model accuracy have a substantial bottom line impact[32], and data scientists owe their clients the best possible predictions.

---

[29]http://www.nytimes.com/2014/08/18/technology/for-big-data-scientists-hurdle-to-insights-is-janitor-work.html

[30]http://www.no-free-lunch.org

[31]http://jmlr.org/papers/v15/delgado14a.html

[32]https://thomaswdinsmore.com/2014/04/09/automated-predictive-modeling/

SAS and IBM recently introduced automated modeling features to their data mining workbenches. In 2010, SAS introduced SAS Rapid Predictive Modeler[33], an add-in to SAS Enterprise Miner. Rapid Predictive Modeler is a set of SAS Macros supporting tasks such as outlier identification, missing value treatment, variable selection, and model selection. The user specifies a data set and response measure; Rapid Predictive Modeler develops and executes a test plan, measuring results from each experiment. The user controls execution time by selecting basic, intermediate, or advanced methods. In 2015, SAS introduced SAS Factory Miner, a more advanced product that runs on top of SAS Enterprise Miner.

IBM SPSS Modeler is a set of automated data preparation features as well as Auto Classifier, Auto Cluster, and Auto Numeric nodes. The automated data preparation features perform such tasks as missing value imputation, outlier handling, date and time preparation, basic value screening, binning, and variable recasting. The three modeling nodes enable the user to specify techniques to be included in the test plan, specify model selection rules, and set limits on model training.

The caret[34] package in open source R is a suite of productivity tools designed to accelerate model specification and tuning. The package includes pre-processing tools for dummy coding, detecting zero variance predictors, and identifying correlated predictors; the package also includes tools to support model training and tuning. The training function in caret currently supports 217 modeling techniques; it also supports parameter optimization within a selected technique, but does not optimize across techniques. Users write R scripts to call the package, run the required training tasks and capture the results.

Auto-WEKA[35] is another open source project for automated machine learning. First released in 2013, Auto-WEKA is a collaborative project driven by four researchers at the University of British Columbia and Freiburg University. Auto-Weka currently supports classification problems only. The software selects a learning algorithm from 39 available algorithms, including 2 ensemble methods, 10 meta-methods, and 27 base classifiers.[36] Since each classifier has many possible parameter settings, the search space is very large; the developers use Bayesian optimization to solve this problem.[37]

---

[33]o14a.html
support.sas.com/resources/papers/proceedings10/113-2010.pdf
[34]-marketing-technologists-are-the-hottest-jo
[35]bs-of-2015/
http://blogs.wsj.com/cio/2014/11/10/for-cios-
[36]https://www.datarobot.com/blog/automated-machine-learning-short-history/
[37]universities-cant-train-data-scientists-fast-enough/
http://www.chica

Challenges in Machine Learning[38] (CHALEARN) is a tax-exempt organization supported by the National Science Foundation and commercial sponsors. CHALEARN organizes the annual AutoML[39] challenge, which seeks to build software that automates machine learning for regression and classification. The most recent conference[40], held in Lille, France in July 2015, included presentations[41] featuring recent developments in automated machine learning, plus a hack-a-thon.

DataRobot, a Boston-based startup founded by insurance industry veterans, offers a machine learning platform that combines built-in expertise with a test-and-learn approach. By expediting the machine learning process, DataRobot enables organizations to markedly improve data scientist productivity and expand the pool of analysts without compromising quality. DataRobot has assembled[42] a team of Kaggle-winning data scientists, whose expertise it leverages to identify new machine learning algorithms, feature engineering techniques, and optimization methods.

The DataRobot platform uses parallel processing to train and evaluate thousands of candidate models in R, Python, H2O, Spark, and XGBoost. It searches through millions of possible combinations of algorithms, pre-processing steps, features, transformations, and tuning parameters to identify the best model for a dataset and prediction problem.

DataRobot leverages[43] the cloud (Amazon Web Services[44]) to provision servers on demand as needed for large-scale experiments; the software is also available for on-premises deployment and in Hadoop. Users interact with the software through a browser-based interface, or through an R API.

---

[38]gotribune.com/business/ct-
[39]indeed-survey-0514-biz-201505
[40]14-story.html
http://www.mckinsey.com/features/big_data
[41]https://hbr.org/2014/09/stop-s
[42]http://venturebeat.com/2015/08/19/this-guy-is-the-superman-of-data-scientists/
[43]http://www.techstars.com/content/blog/sponsorship-in-action-datarobot-and-aws/
[44]https://medium.com/aws-activate-startup-blog/datarobot-leverages-aws-for-predictive-modeling-dae62ba2a523#.xrwen5lxd

In August 2014, DataRobot raised[45] $21 million in Series A venture capital financing. Recruit Holdings, a Tokyo-based company, announced[46] an investment in DataRobot in November 2015. DataRobot announced an additional $33M in a Series B round on February 11, 2016.

The New Self-Service Analytics

In this chapter, we surveyed six key innovations in self-service analytics:

- Self-service visualization via Tableau and its imitators.

- Self-service data blending via Alteryx and similar products.

- BI in Hadoop, especially middleware like AtScale that enables organizations to leverage existing BI assets.

- Cloud-based prebuilt services like Domo that enable functional managers to bypass the IT bottleneck.

- Business-oriented open core analytics platforms like RapidMiner and KNIME that enable collaboration between experts and business users.

- Open and transparent expert systems for machine learning like DataRobot that make machine learning accessible for a broader pool of users.

Self-service visualization tools like Tableau work most effectively with single tables of clean data, since they lack strong data blending capabilities. Consequently, they belong at the end of an analytics value chain, where they facilitate collaboration between expert and non-expert users. End users working with Tableau, for example, can visualize data in many different ways; this saves enormous amounts of time for expert analysts, who can deliver a table or dataset rather than hundreds of charts.

The combination of a data blending tool like Alteryx and a data visualization tool like Tableau offers a powerful set of self-service capabilities. Complex data blending with rough data sources requires a relatively high level of skill in Alteryx, so this combination is better suited to the business analyst than the casual information user.

---

[45]http://blogs.wsj.com/venturecapital/2014/08/15/datarobot-run-by-worlds-top-data-scientists-raises-21m-series-a/
[46]http://www.recruit-rgf.com/news_data/release/2015/1117_7730.html

Enterprises should strive for "BI everywhere"—the idea that an end user should be able to use the same self-service tooling regardless of where the data is physically stored. Tableau partially accomplishes this end because it has a flexible and easily configured back-end that can work with a wide range of data storage options. However, pointing Tableau directly at a Hive metastore in Hadoop isn't for the faint-hearted. AtScale middleware makes a Hadoop cluster as easy to access as a relational database.

Insight-as-a-service offerings like Domo seek to support the *entire* analytics value chain. Their value proposition is speed and simplicity for the functional manager; with predefined reports and dashboards, they can quickly deliver essential information, largely bypassing the IT organization altogether.

Analytics platforms like RapidMiner and KNIME inhabit a middle ground between analytic programming languages (such as Python and R) and simple desktop analytic tools. With a workflow-oriented drag-and-drop interface, they save time for the business analyst, while offering rich analytic functionality. They are also highly extensible, offering the ability to embed user-defined functions in a workflow.

Automated machine learning tools like DataRobot save enormous amounts of time for expert data scientists, and they also broaden the pool of people in an organization who can build predictive models. With the ability to tightly integrate with production systems, they radically reduce the time to value for machine learning.

People in organizations have diverse needs for analytics; there is no single tool that meets all needs. Enterprises will continue to employ experts even as they invest in simple tools with broad appeal. If anything, the job market for experts is tighter than ever.

Handbook for Managers

How to Profit from Disruption

Let's review what we have covered so far in this book.

Chapter One defined disruptive innovation and makes the argument for disruption in the business analytics marketplace today. The remainder of the chapter covered basic concepts, such as the demand for data-driven insight and the characteristics of the analytics value chain. We distinguished between disruption within the analytics value chain, and disruption of other markets by analytics.

Chapter Two briefly recapped the history of business analytics in the modern era. We showed previous examples of disruption in the value chain, such as the introduction of the enterprise data warehouse. We also provided examples where analytics disrupted other markets, as in the cases of credit scoring and fraud detection. Finally, we introduced you to key trends driving disruption today, including the digital transformation of the economy and declining costs of computing and storage.

Chapter Three detailed the open source business model. We introduced you to open source licensing and distribution, and to commercial business models based on open source software. We also provided a detailed profile of Python and R, the two leading open source software platforms for analytics.

© Thomas W. Dinsmore 2016
T. W. Dinsmore, *Disruptive Analytics*, DOI 10.1007/978-1-4842-1311-7_10

In Chapter Four, we covered Hadoop and its ecosystem. We noted the distinction between Apache Hadoop and its commercial distributions, and documented the components most widely used together with Hadoop. Under analytics, we distinguished between the rudimentary capabilities available under Hadoop 1.0 and the increasingly powerful and sophisticated capabilities available today, under Hadoop 2.0.

Chapter Five documented the rapidly declining cost of computer memory and the corresponding rise of large-scale in-memory computing. This chapter covers Apache Spark, Apache Arrow, Alluxio (Tachyon), and Apache Ignite.

In Chapter Six, we briefly surveyed the history of streaming analytics, taking note of the longstanding gap between vision and reality in the category. We surveyed streaming data sources, such as Apache Kafka and Amazon Kinesis, and open source streaming analytics platforms: Apache Apex, Apache Flink, Apache Samza, Apache Spark Streaming, and Apache Storm.

Chapter Seven covered fundamentals of cloud computing and the elastic business model. We surveyed the capabilities of the top three cloud platforms: Amazon Web Services, Microsoft Azure, and Google Cloud Platform.

In Chapter Eight, we reviewed key trends in machine learning: convergence, competitions, ensemble learning, scalability, and Deep Learning. We offered a detailed introduction to neural networks and Deep Learning. Finally, we surveyed advanced machine learning platforms, including distributed engines, in-database libraries, and Deep Learning frameworks.

Finally, in Chapter Nine, under self-service analytics, we offered a balanced perspective on the role of casual and expert users in enterprise analytics and proposed a model of user personas. We profiled a number of products that exemplify innovations in this area, including data visualization, data blending, BI on Hadoop, "Insight as a Service," business user analytics, and automated machine learning.

At this point, if you're not convinced of the power and richness of emerging innovations in analytics today, stop reading; Chapter Ten is not for you.

In this chapter, we offer the manager a handbook for action to profit from innovation and disruption. We cover three broad areas:

- People and organization

- Processes

- Platforms and tools

Some of the strategies we propose may seem radical. This handbook assumes that you want your organization to profit from innovation and disruption in the analytics marketplace. Either you do or you don't—only you can decide.

To profit from innovation and disruption, it's likely that your organization will need to do some things differently. There may be political or other barriers to overcome, and change management is always a concern. We don't trivialize the difficulty of organizational change. However, this is not a book on organization politics or change management; it's about how to profit from innovation and disruption.

People and Organization

In analytics, we tend to focus too much attention on purely technical problems. But people and organization matter—especially so in a disrupted world.

- Organize around clients.
- Define the Chief Analytics Officer's Role.
- Make costs visible to clients.
- Hire the right people.

Motivated, well-organized people with basic tools and the right incentives outperform poorly motivated and poorly organized people with gold-plated tools every time.

Organize Around Clients

Organizing for a disrupted world requires a laser-like focus on client needs. (We use the term *client* instead of *customer* because the practice of analytics is a professional service. Clients can be internal or external — the same principles apply.) Every organization is different; we present here a typical model of analytic needs together with the skills and tools needed to meet those needs.

Enterprises typically organize analytics around technical functions: data integration, data warehousing, business intelligence, and so forth. There is a logic to this, including knowledge sharing and career development. But it is imperative that analytics teams organize around internal and external customers.

In Chapter One, we outlined five distinctly different sources of demand for data-driven insight:

- **Strategic**: Insight for C-level executives.
- **Managerial**: Insight for functional executives.
- **Operational**: Insight for business process optimization.
- **Developmental**: insight for new products and services.
- **Differentiating**: insight for (external) customers.

Each of these groups of internal and external clients has distinct needs for people, skills and tools:

Strategic. Most of the analysis for C-level executives is ad hoc, unrepeatable, and urgent. Practitioners require deep knowledge of the business or industry, a highly professional approach, and a strong grasp of visual presentation techniques. A strategic analysis team requires broad access to internal and external data, as well as capabilities for ad hoc data integration, queries, and reporting.

Managerial. Rigorous performance measurement and ad hoc analysis for business planning are the principal requirements at this level. Practitioners often have a finance background; they must be familiar with the organization's performance metrics and business planning process. Conventional data warehousing and business intelligence systems perform well for performance metrics. A managerial analytics team must be able to access the performance measurement system for ad hoc performance reports and needs tools to develop forecasts for business plans.

Operational. Business process stakeholders need low-latency real-time metrics of the business process, and they need deployable machine learning tools for optimization. Conventional business intelligence and reporting systems work well for operational metrics if they are deployed for real-time analysis. Operational analysts should have a strong background in statistics, machine learning, and content analytics, with the programming skills needed for model deployment.

Developmental. Product and service development executives need insight to support the development lifecycle from concept through product introduction. Analytics practitioners need a background in consumer and marketing research combined with statistical training in experimental design, "test and learn" techniques, and forecasting. Lightweight "desktop" tooling is usually sufficient in this area, since these practitioners rarely work with Big Data.

Differentiating. Customer facing analytics, such as recommendation engines, are necessarily production-grade applications. Open source software is preferable to commercial software, especially if the organization plans to distribute software components to the end customer. Practitioners in this area should have a software engineering background supplemented with machine learning training. Knowledge of programming languages—such as C, Java, Scala, Python, and R—is required.

While there is a great deal of variation across organizations, the general rule is that junior-level analysts tend to report to the departments they support, such as marketing or credit risk; expert analysts tend to be centrally grouped; and specialists in data integration, data management, software administration, and provisioning tend to report to the IT organization. As a result, no senior executive holds responsibility and accountability for the analytics value chain as a whole.

A better organization model separates technical functions in the analytics value chain from IT and places them under a Chief Analytics Officer (CAO). Working analysts either remain in the functional organizations with a dotted line reporting relationship to the CAO or they report into the CAO organization and are assigned to support functional organizations.

Define the Chief Analytics Officer's Role

There is an emerging trend towards designating an executive as the Chief Analytics Officer (CAO). While this is not yet a universal practice, it signals that the enterprise believes analytics is a key strategic capability.

In theory, the Chief Analytics Officer (CAO) should be accountable for the entire analytics value chain, from data to insight. In practice, however, responsibility for the analytics value chain tends to be divided and is likely to remain so. The IT organization usually manages the data warehouse, the processes that acquire data, and "enterprise" grade business intelligence tools. IT also manages hardware and software procurement.

Functional departments manage "shadow" IT operations, which may include data marts and analytics tooling. "Expert" users sometimes reside in IT; more often, they reside in functional departments. In specialized analytic disciplines, such as actuarial analysis or credit risk, analysts generally report to a functional manager; this is unlikely to change.

Divided responsibilities lead to some dysfunctional outcomes. Since IT generally owns the data warehouse but not the delivery of insight, there is a tendency to view the collection and management of data as an end in itself; actual insight is someone else's concern. While IT organizations are often very capable in the construction and management of data warehouses, they tend to overlook or ignore functional managers' *unmet* needs. That is why so many functional managers have their own "shadow" IT operation.

The CAO should directly manage the team responsible for strategic analytics (as defined in the previous section). This team handles ad hoc requests for insight from the organization's leadership, and it should be staffed and tooled accordingly. A strategic analytics team requires highly skilled and professional people with broad access to internal and external data and the tooling necessary for quick response.

For managerial analytics, the principal asset is the performance measurement system: the enterprise data warehouse and supporting business intelligence platform. IT's core competence is in the operation of production systems, so it makes sense that the CIO manages the performance measurement system. The CAO, however, should own the process for driving requirements for performance measurement, coordinating across functional stakeholders.

Business process optimization typically requires skills in advanced analytics and operations research. Some functional departments may already have expert teams to support these capabilities; however, it also makes sense to pool these resources centrally to support departments that have not yet developed their own team. The CAO should manage these pooled resources, and also drive standards and best practices across the organization.

Most organizations have dedicated teams for product and service development, typically domiciled in the Marketing organization. The CAO's role in this process should be to drive training and adoption of best practices, broker requirements with the IT organization, and define common standards.

Software and hardware provisioning require careful balancing of the CAO and CIO responsibilities. The CIO generally manages on-premises provisioning and often manages cloud provisioning as well. Consistent with security standards, however, the CAO should be free to move workloads to the cloud if the organization cannot provide competitive pricing or service levels.

In a similar manner, while the CIO generally manages software licensing, procurement, and support, the CAO should own this responsibility for advanced analytics software, which generally falls outside of IT's core competence.

Data ownership is another area that requires careful balancing of CIO, CAO, and functional responsibilities. Data ownership and management are two different things. The data owner controls data access (within the framework of an organization's overall policies) and defines the business rules under which the data is captured; the data manager handles administration and custody on behalf of the owner. Generally, data should "belong" to the organization that funds its production; the CIO should manage production systems and databases; the CAO should manage analytic datastores.

Make Costs Visible to Clients

The question of "chargebacks" may seem overly detailed for a book on analytics, but incentives matter.

The author once met with an IT executive of a large healthcare provider, who expressed frustration that SAS users were unwilling to switch to lower-cost alternatives. Asked if user departments contributed to the cost of the software, the executive replied that no, the organization wanted to encourage people to use analytics.

You can see the problem there: when costs are invisible, users prefer the gold-plated option.

There are only two viable models for software provisioning. Either software selection is a centrally managed process, with costs absorbed as overhead, or individual departments and users can pick and choose their software and pay the costs out of their own budgets.

The same principle applies to people costs, provisioning, and data collection. Nothing clarifies needs so quickly as a requirement to pay for what you use.

Hire the Right People

In mid-2016, there is a seller's market for qualified data scientists. As a result, many people call themselves data scientists who aren't really qualified. In the absence of professional certification, there are many articles in the media[1,2] covering "interview questions" for data scientists. Most of these cover data science trivia, and many highly accomplished analysts would fail.

Testing candidates on theoretical knowledge is a misguided approach. A better approach focuses on asking the candidate about actual projects they have completed:

- What business problem did you address?

- How did you go about solving the problem?

- Did you work with others? If so, describe your interactions with them?

- What tools did you use? What technical problems did you solve?

- In the course of the project, what worked well? What could have worked better?

- How did the project end?

Top candidates can easily point to dozens of projects to which they have contributed. Many excellent data scientists are active in Kaggle or other competitive platforms and may have contributed to an open source analytics project; these are indicators that the individual has a good command of the discipline.

At a junior level, the key characteristic to assess is motivation: does this person have a burning desire to perform analysis? Even new graduates can point to examples of analysis they have done—research projects, or independent work with public data sets. Learning an open source analytic language like R or Python is another key indicator of motivation.

A candidate for a position in analytics who cannot point to actual projects, or who has never learned an analytic language, is not serious.

---

[1]https://www.dezyre.com/article/100-data-science-interview-questions-and-answers-general-for-2016/184
[2]http://www.datasciencecentral.com/profiles/blogs/66-job-interview-questions-for-data-scientists

Analytics executives sometimes ask if candidates should be required to know specific analytic tools and, if so, which ones. The short answer to the question is that it depends on what platform you have established as a standard. (If you haven't established a standard, you need to do so. See "Build an Open Source Stack" later in this chapter). If you have standardized on SAS, look for people who know SAS. While it is theoretically possible to retrain a candidate who is otherwise qualified, some individuals never make the transition.

For any candidate in any organization, cultural fit is essential. It is unwise to generalize about the personalities of data scientists or analysts; it's fair to say, however, that successful data scientists and analysts are able to engage with clients to understand business problems, explain results, and work collaboratively on a team. Gauging these qualities should be part of your evaluation.

Processes

Principles of agile development, as expressed[3] in the Agile Manifesto, apply to business analytics as well as to general software development. For convenience, we restate the 12 principles (slightly paraphrased for business analytics):

1. Satisfy clients through early delivery.
2. Welcome changing client requirements.
3. Deliver work product frequently.
4. Cooperate closely with business stakeholders.
5. Build motivated teams, and trust them.
6. Communicate face-to-face.
7. Work product is the principal measure of progress.
8. Work at a sustainable pace.
9. Focus continuously on technical excellence and good design.
10. Simplify problems.
11. Self-organizing teams deliver the best architectures, requirements and designs.
12. Reflect regularly on how to be effective and adjust accordingly.

---

[3]http://www.agilemanifesto.org/principles.html

We apply these principles separately to business intelligence and machine learning in this chapter.

Separately, we note that IT-led data warehousing operations often collect too much of the wrong kind of data, and not enough of the data needed to drive critical insight. To correct that, we propose a lean data strategy.

Practice Agile Business Intelligence

Despite the growth of self-service BI, most organizations continue to employ trained specialists to satisfy ad hoc requests for analysis. These specialists are often staffed centrally; functional teams request analysis through written requests and detailed requirements. Specialists get into the habit of delivering exactly what is spelled out in the requirements. Since functional managers lack the expertise of the specialists, they may not know precisely what they want; this communications problem leads to disputes. The specialist team always has a work backlog, so any request takes days or weeks, no matter how trivial.

While self-service BI mitigates this problem by engaging the requestor directly in production of the analysis, there are limits to what can be accomplished simply through tooling. Self-service BI works best when the data is well-organized, accessible, and limited in scope. Even with the best BI tools, managers tend to delegate the BI task, so self-service BI may simply create a new breed of specialist.

Agile principles suggest that organizations can resolve the BI bottleneck effectively simply by distributing specialists into the functional teams, co-locating them for maximum interaction with business stakeholders. This approach makes it possible for specialists to anticipate business needs for insight, help the requestor frame the requirements, and work interactively with the requestor to develop a solution.

This approach does not rule out investing in self-service BI tools. As a rule, organizations that distribute BI specialists into functional teams discover that the total demand for data-driven insight increases. The BI specialist serves to spearhead broader use of the self-service tool and collaborates with business stakeholders when self-service tools are not sufficient to solve the problem.

Practice Agile Machine Learning

Machine learning differs from business intelligence for two reasons: the deliverable is a working predictive model rather than a report, table, or chart; and because the process itself requires a higher level of skill and expertise. Another key difference: while functional teams constantly use business intelligence, the demand for machine learning tends to be more selective. In the typology of demand for insight discussed earlier in the chapter, the greatest demand for machine learning stems from operational business process optimization and from differentiating products and services.

Rather than domiciling expert data scientists in specialist teams for short-term project engagement, agile principles suggest that organizations will achieve better results by placing data scientists directly into process improvement or product development teams. As is the case with BI specialists, placing data scientists into teams ensures a collaborative approach to the design, development, and evaluation of machine learning models. It also enables the data scientist to develop domain knowledge and an understanding of the business context for machine learning.

Agile principles imply some changes to standard data science practices.

First, the work product from machine learning is a production scoring model. This differs from standard practice, where modelers often view their work as complete once they build a satisfactory model in the lab environment; deployment is someone else's problem.

Second, data scientists should evaluate predictive models solely on how they perform in production. "Sandbox" testing is useful for preliminary model selection, but performance in production is not always the same as sandbox performance. Where they differ, production performance is the correct metric.

These two principles imply a more rapid cycle time into production than data scientists may be accustomed to. It implies an approach where the data scientist seeks to quickly deliver an unbiased model that outperforms naïve criteria, then continuously improve it by examining prediction errors, supplementing data sources, testing new training algorithms, and so forth.

The need for rapid deployment implies a third key principle: as much as possible, data scientists should avoid modifications to the production data that cannot be reproduced in a scoring model. Data scientists like to enhance raw data in various ways that improve the predictive power of a machine learning model. But these modifications can make it more difficult to deploy the model, since any changes to the data in the modeling process must be reproduced in production.

Rapid cycle time also has implications for tooling. Platforms that automate routine data science functions and enable large-scale testing are highly desirable. So are capabilities that tightly integrate model development with model scoring.

Develop a Lean Data Strategy

In 2015, technology consultant Forrester surveyed[4] more than 1,800 technology decision-makers in organizations around the world. Forrester asked respondents to estimate the percentage of the data currently used for business

[4]http://www.forrester.com/pimages/rws/reprints/document/116447/oid/1-SFDMEH

intelligence. Respondents reported separately for structured, semi-structured, and "unstructured" data; the average response by category was:

- Structured data: 40% used

- Semi-structured data: 27% used

- "Unstructured" data: 31% used

Forrester interprets this low utilization as a problem with tools: if organizations simply invest in self-service business intelligence tools, end users will tap the data. *There is golden insight hidden inside that unused data; all your organization needs is to buy another piece of software and all will be revealed.* However, there are a number of possible explanations for the low data utilization other than tools availability:

- Data may be structured and stored in a schema that is difficult for most users to navigate, or in a relational model that does not match the way managers think about the business.

- Data lineage and metadata may be poorly documented, so that managers do not trust the data.

- Data security policies may be unduly restrictive, preventing wide use of the data in the organization.

- Data may not be catalogued, and prospective users simply do not know what data is available.

Note that if any of these conditions are true, the IT organization's data warehousing initiative has failed. There may be good reasons for the failure, such as lack of budget, resources, or strategic alignment. But failure is failure.

There is one more possibility: the data is not used because it has no useful information value.

That idea may seem heretical in the era of Big Data, but let's take a moment to explore it. By definition, *data that nobody uses has no value.* (It may have *potential* value for some theoretical future user, but until that user materializes the data is just sitting around taking up storage space.) The whole point of collecting and managing data is to produce useful insight; no insight, no value delivered. The only question that matters is whether the unused data has *potential* value; is there gold inside that pile of junk, or is it just junk?

There are a number of reasons to be skeptical of claims that your unused data has valuable "hidden" insight:

- Any data—whether it is structured, semi-structured, or "unstructured"—is accessible with the right tools tools; if your unused data is valuable, why isn't anyone using it today?

- Motivated analysts climb mountains to get valuable data; if necessary, they learn new tools. Are your analysts unmotivated? New tools won't solve that problem.

- There are few recorded cases (if any) where an analyst produced useful insights by trolling through "found" data.

On that last point, data warehousing vendors have hyped the value of such trolling for years. The best example is the "beer and diapers" story.

In 1992, an analysis team at Teradata analyzed 1.2 million market baskets from 25 Osco drug stores and discovered that between 5:00 p.m. and 7:00 p.m. customers purchased beer and diapers together.[5] Osco never did anything with the insight, because there were no clear merchandising implications. Nevertheless, Teradata's marketing team cited the example as the kind of insight that justifies investing in a data warehouse. The beer-and-diapers story became part of the folklore of the data mining community.

In the same Forrester survey, two-thirds of the respondents reported that the majority of their organization's business intelligence needs are met by "shadow" IT operations—processes that functional managers assemble by themselves. Managers do not sit passively and wait for IT to deliver the intelligence they need; they actively build their own processes, hiring people and investing in tools if necessary to do so.

To summarize: organizations collect a lot of data that is not used. At the same time, they fail to deliver the information functional managers *do* want to use. That is dysfunctional.

There are several reasons for the dysfunction.

One is a lack of input from users and prospective users. It seems obvious that user input is essential to good data warehouse design; yet, anyone with working experience in enterprise analytics can cite examples of highly touted projects built without it. Collecting user input is hard, and it takes time; prospective users often do not know what they want or need, and may not have stable information needs.

Another is a sort of inertia—in the absence of clear design, it is easier to simply copy data from a data source into a data warehouse structure and leave it at that. The author once worked with a global consumer marketer that maintained two data warehouses: one fed exclusively by its SAP ERP system, and the other fed exclusively by its Oracle CRM system. Users who needed data from both systems downloaded summary data and performed the consolidation in spreadsheets.

---

[5]http://www.dssresources.com/newsletters/66.php

A third reason is a phenomenon best described as *data fetishism*, a belief in the magical powers of data, where more data is always better than less data or no data. The problem with this sort of thinking is that data is not a commodity, like crude oil or pork bellies, any unit of which is substitutable for any other unit. To the contrary, data is always particular to a specific event or set of events; a piece of data either answers a question or it doesn't. Petabytes of data are worthless if they do not answer a question.

Cheap storage also encourages organizations to "squirrel" away data whose value is unclear. The cost per terabyte of disk storage has declined precipitously in the past decade, continuing a long-term decline in all computing costs. But while storage is cheap, it is not free; and while the cost of physical storage is declining, the costs of data governance, management, and security are not.

The term *data warehouse* is a metaphor borrowed from logistics, where the purpose of a warehouse is to store inventory. Imagine a warehouse for a retail chain where half the goods are unwanted, while store managers scramble to avoid stockouts by procuring the goods they need through other channels. The warehouse metaphor is doubly ironic when you consider that for the past 20 years and more, enterprises have gone to great lengths to reduce or eliminate inventories through lean manufacturing and just-in-time logistics.

What is the data warehousing equivalent of lean manufacturing?

First, do not acquire data unless there is a clear business need for the information it carries. In practical terms, "business need" means that a functional manager with a budget is willing to pay for the data. Stop acquiring data when the business need ends.

Second, build metrics into products, processes, and programs from inception. Do not create performance metrics after the fact; design them into every business entity. Include the cost of performance metrics into product and program financials.

Third, align data presentation and user personas. Typically, early adopters for a particular set of data are expert users who can work with messy and granular data, developing insights on behalf of business stakeholders. If and when business requirements stabilize, the analytics developed by experts can be productionized and made accessible to users who prefer to work with simpler tools.

Fourth, do not "clean" data; data cleansing tools do not make data more accurate; they simply make it *appear* more accurate by removing anomalies. Anomalies, however, have information value; Alan Turing and colleagues at Bletchley Park used them to decrypt the Enigma cypher. If a data source systematically produces erroneous data, fix the data source.

Finally, do not make the data warehouse an end in itself. At all times, the goal should be delivering insight; development initiatives should organize around specific projects to deliver insight to specific individuals, teams, functions, or applications. It may be possible to identify common data consolidation needs across multiple end user applications; when that is the case, a data warehouse can serve as an omnibus platform across these applications. But if it is difficult to define such commonalities, do not let the data warehouse idea get in the way of delivering insight.

Platforms and Tools

Unless your organization is a startup, you have a legacy tools environment—existing investments in tools to support various elements of the analytics value chain. Your organization's past investments in software and hardware are a sunk cost; in many cases, it is more cost-effective to retain existing tools than it is to replace them. We are not suggesting that you toss out existing tools if they still meet your needs.

On the margin, however, there are things you can do to profit from disruption. First, assess what you are actually using and match this to your licensing; for incremental expansion, define needs rigorously. Second, build a credible open source alternative; even if you don't use the open source option heavily, simply having it gives you negotiating leverage with commercial software vendors. Finally, leverage elastic provisioning—in the cloud, or through on-premises virtualization.

Assess Software Licensing and Use

Well-defined requirements are essential in a disrupted world because the conventions we use to anchor decisions don't work anymore. Industry leaders struggle; outsiders bring new capabilities to market; established experts struggle to adapt. Organizations that know what they actually *need* and what they don't need thrive in this environment.

"Define your needs" seems like obvious guidance, but it is surprising how often one encounters analytics managers who have only a cursory understanding of how their team uses tools. Formal assessment usually reveals that people use only a fraction of the functionality embedded in commercial software tools. Now, more than ever, you need to take a cold, hard look at your commercially licensed software and how it is used.

Open source software disrupts commercial software by delivering "good enough" functionality under a services-based business model. Commercial software vendors point out that their software products have more features than open source software. This is accurate, but misleading. Features only add value when your organization actually uses them; otherwise, they simply add cost.

If you do not have well-defined requirements, software selection will gravitate to the products with the most features, the best marketing, the best analyst relations, or all three. Rather than selecting software based on which one has the most features, choose the lowest-cost product that satisfies all of your organization's demonstrated needs.

A side benefit of such an assessment: your organization is almost certainly overlicensed. The software industry focuses on underlicensing and pirated software; but unless your organization has actively managed software licensing, the odds are that a sizeable share of your total software spending goes to shelfware.

How do we know that you are overlicensed? Because commercial software vendors fear that elastic "pay for what you use" pricing will cannibalize their existing software licensing models. That fear is justified; while vendors like Oracle, SAP, IBM, and Microsoft all report double-digit growth in cloud-based revenue, that growth fails to offset the decline in conventional software licensing revenue.

If software vendors get less revenue from elastic "pay for what you use" pricing, it follows that the standard commercial licensing model makes you pay for what you do not use. This makes some sense when you consider commercial licensing terms, which require the buyer to pay in advance for the right to use software whose business value is unknown.

Commercial vendors warrant that software does what they say it does, and that it works under specified conditions. However, commercial vendors *sell* software based on business value and not on features and functions. They do not warrant these claims of business value; the buyer assumes this risk.

When your organization has a well-defined set of requirements for business analytics software, you are in a much better position to evaluate the claims of commercial vendors against one another and against an open source stack.

Build an Open Source Stack

Across the software economy, the open source business model is undermining the commercial software model. Consultant IDC writes[6]:

> Open source products offer functionality that is competitive with proprietary products and applies downward pricing pressure on these products. Growth in the adoption of open source technologies will force an acceleration toward a services-based business model for many vendors.

---

[6]https://www.idc.com/getdoc.jsp?containerId=257402

There are open source software alternatives for every component in the analytics value chain:

- Hadoop and its ecosystem offer comprehensive tooling for data acquisition and management.

- Open source SQL engines—such as Spark SQL, Impala, Drill, and Presto—compete successfully against data warehouse appliances for interactive queries.

- Machine learning engines like H2O and Spark MLlib provide scaleable machine learning options. R and Python are excellent general-purpose platforms for analytics.

- JasperSoft, Pentaho, and Talend all deliver end-to-end capabilities for business analytics.

To build and deliver an open source stack, follow these steps:

- Establish an analytics innovations team.

- Assess your organization's current open source software usage.

- Evaluate open source components through live testing and pilot projects.

- Define a support strategy.

An innovations team is a core group of individuals whose primary role is to evaluate innovative technologies and bring them into the organization. Members of the team may be full-time, or may be temporarily assigned from other roles; however, the team's impact and time to value depend on its leadership, personnel, and resources.

Once your team is established, take stock of your organization's current use of open source software. The results of such an assessment may surprise you. Many business leaders simply do not know the extent of open source software use in their organizations, because there is often no central control over acquisition and use of open source software.

The open source software your organization used successfully forms the foundation of your stack. From there, your team can define additional components to support functional gaps in the stack. Nobody can define the perfect open source stack for your organization; it depends on your needs, your previous experience, and the results of your ongoing evaluation.

Defining a support strategy is essential for your open source stack. There are two aspects to this problem:

- Support for your end users
- Support for your help desk

One way to mitigate the need for support is to choose supported open source distributions. Cloudera, Hortonworks, and MapR offer commercially supported bundles built on Apache Hadoop; Microsoft and Oracle offer supported R distributions; JasperSoft, Pentaho, and Talend all offer commercially supported versions of their products.

However, there is no single open source or open core product that comprehensively supports the entire analytics value chain. Consequently, your help desk plays a key role in diagnosing issues and directing them to the appropriate source for support.

Your open source stack serves as a baseline architecture. This does not mean that you will never use commercially licensed software; it means that you will use commercial software only when the open source stack lacks features and functions that are needed to solve a specific business problem.

A credible open source stack also creates negotiating leverage with commercial software vendors, who deeply discount their software when competing with an open source alternative. Consider the full software lifecycle when evaluating these discounts; some vendors simply discount the first year subscription fee or discount a perpetual license while increasing maintenance fees.

Leverage Elastic Provisioning

Once you have defined an open source software stack, you need to provide computing infrastructure, such as servers and storage. Choose an elastic solution: public cloud, virtual private cloud, private cloud, hybrid cloud, or on-premises data center virtualization and cluster management tools.

Elastic provisioning means that the computing resources available to users expand and contract based on actual workload. For example, if a user needs to run complex analysis to support a prospective merger, the computing resources expand accordingly; when that project is completed, the user releases the resources for use by other applications. *Self-service* provisioning means that end users can requisition additional resources without IT support or intervention.

Three key principles should govern your organization's approach to elastic computing:

- The computing infrastructure that you own and manage for business analytics should operate at a high level of utilization.

- Computing resources for end users should be delivered through self-service elastic provisioning.

- Computing costs (internal or external) should be metered and charged to the consuming application.

The breakeven capacity threshold for your organization depends on the efficiency and skill of your data center team and your level of skill in procurement. Cloud data centers operate at about 65% of capacity; the average utilization[7] of on-premises servers is in the range of 12-18%, so most organizations have a lot of room for improvement.

As noted in Chapter Seven, average infrastructure utilization is low because organizations provision to support peak demand; during periods of slack demand, this computing capacity sits idle. Imagine a company with analytics teams in New York and Singapore, each with a dedicated server. Each team uses its server actively during local business hours, but each server sits idle outside of business hours. This company can double its server utilization and cut computing costs in half if the two teams can share computing infrastructure.

To optimize provisioning, segment your analytic workloads into three categories:

- *Baseline* workload is predictable at a certain constant level.

- *Peak* workload is predictable at a higher level than the baseline for short periods. Month-end reporting, for example, typically creates a short-term spike in demand.

- *Surge* workload is an unpredictable spike in demand above the baseline level. For example, when an analyst trains a Deep Learning algorithm.

Under this framework, provision your baseline workload with infrastructure that you own and manage and can operate at a high percentage of capacity. For peak workload, use reserved instances; for surge workload use on-demand or spot cloud. Of course, you won't want to move data back and forth from the cloud, so you should group workloads together that use common data.

---

[7]https://aws.amazon.com/blogs/aws/cloud-computing-server-utilization-the-environment/

It's entirely possible that your analysis will show a very low baseline workload for analytics. That's typical; workloads for an analytics platform tend to be inherently variable and difficult to predict, because much of the demand is ad hoc and project oriented. Nevertheless, computing and storage must be sufficient for high performance on large-scale problems.

If your workloads are mostly unpredictable, or if your organization lacks the skills to manage computing infrastructure effectively, put everything into the cloud.

Executives tend to raise three objections to the use of off-premises cloud computing: out-of-pocket costs, concerns about outages, and security concerns.

Cost concerns about cloud are largely an illusion. Cloud computing costs are measurable and tangible, while internal computing costs are often hidden away in depreciation charges, salaries, floor space, and electric bills. Even if the organization is very good at measuring costs and charging back costs to users, it still pays for unused capacity. With their purchasing efficiencies and skilled data center management, cloud data centers achieve economics of scale that most organizations can only dream about.

While anxiety about data center outages is real, there is no evidence that cloud data centers are more vulnerable to outages than on-premises data centers. Any data center is subject to outages for any number of reasons: natural disasters, cyber attack, or human error. As with cost concerns, there may be an illusory sense of security in an on-premises facility: of course, *our* people won't mess up and bring the system down. Keep in mind that many organizations run mission-critical applications in the cloud today—and business analytics applications are rarely mission critical.

Security concerns are similar to concerns about outages: the anxiety is real, but there is no evidence that cloud data centers are less secure than on-premises data centers. (If anything, the opposite is true: in the past two years, 19 of the top 20 data breaches hit on-premises data centers; the one breach of cloud data is fully attributable to human error, and not related to the physical location of the data.)

In any case, there are work practices analysts can implement to minimize security risks in the cloud. These include avoiding use of Personally Identifiable Information (PII), which is sensitive information about individuals that is rarely needed for analysis; removing identifiers from table names and column headers; and the use of hashing to encrypt data before it's transferred to the cloud.

Elastic self-service provisioning with metered costs should be the standard of service provided to users. This is the service standard delivered by cloud providers; if your organization cannot stomach using off-premises cloud platforms, your goal should be to deliver the same level of service through data center virtualization. This is an essential requirement for an analytics platform.

Closing Thoughts

Your perspective on disruption depends on where you stand.

- If your organization buys and uses analytic software and services, disruptive innovation is an opportunity for you to improve the effectiveness of your investments in analytics and to reduce costs. Avoid getting locked into vendors who are ripe for disruption.

- If your organization seeks to disrupt others with innovative products and services, the open source projects described in previous chapters offer an excellent foundation.

- If your organization has an established franchise providing business analytics software and services, *watch your back*; someone out there wants to eat your lunch.

To profit from disruptive innovation, do the following things:

- **Organize Around Client Needs for Data-Driven Insight**. Stop thinking about analytics as a single problem that some big vendor can solve for you. Your clients have diverse needs for data-driven insight; tailor solutions accordingly.

- **Carefully Define the Role of the CAO**. In most cases, it is impractical to expect any single executive to "own" the complete analytics value chain. Assign the CAO accountability to drive data-driven insight in the organization, then carefully balance roles and responsibilities of the CAO, CIO, and functional executives.

- **Align Decision-Making Authority Over Analytics Platforms with Responsibility for Costs**. Avoid scenarios where users choose platforms without cost accountability, and costs are "someone else's problem".

- **Hire the Right People.** For analysts, place less emphasis on credentials and theoretical knowledge, and more emphasis on analytic accomplishments and collaboration skills.

- **Practice Agile Business Intelligence**. Deploy specialists to functional teams and encourage close collaboration. Invest in self-service tools if there is a demand, but don't assume that your need for analytic specialists will go away.

- **Practice Agile Machine Learning**. Focus on repeatable processes, reduced cycle time, rapid deployment, and continuous improvement of the production model. Invest in platforms that maximize the productivity of your high-value data scientists.

- **Develop a Lean Data Strategy**. Stop thinking of your data warehouse as a strategic investment; it's not. Align data collection with needs for data-driven insight. Do not collect data for which there are no defined users and stakeholders.

- **Assess your Commercial Software Licensing and Usage**. Take a cold, hard look at software licensing in your organization; you are almost certainly overlicensed today. Challenge those who insist they must use high-end commercial software.

- **Build an Open Source Stack**. Define, build, deliver, and support an open source stack for business analytics. Make the open source stack your baseline system. Use commercial software only when your organization's documented requirements can't be met with your open source stack.

- **Leverage Elastic Provisioning**. Analytic workloads tend to be ad hoc and unpredictable, which makes them excellent candidates for elastic provisioning—in the cloud or on-premises.

The payoff for taking action: more effective analysts, more data-driven insight, better decisions, and lower total cost of ownership for your analytics infrastructure.

The penalty for inertia won't be visible right away. Your business analytics software vendors will continue to send you renewal invoices. The cost of decisions *not* taken, of data-driven insights *not* produced, will never be measured. Life will go on.

At some point, however, someone will ask: "why are you here?"

Some people in your organization may object to the measures we outline in this chapter. They may even call them disruptive.

If they do, smile. You're on the right track.

Index

I

Get the eBook for only $5!

Why limit yourself?

Now you can take the weightless companion with you wherever you go and access your content on your PC, phone, tablet, or reader.

Since you've purchased this print book, we're happy to offer you the eBook in all 3 formats for just $5.

Convenient and fully searchable, the PDF version enables you to easily find and copy code—or perform examples by quickly toggling between instructions and applications. The MOBI format is ideal for your Kindle, while the ePUB can be utilized on a variety of mobile devices.

To learn more, go to www.apress.com/companion or contact support@apress.com.

Printed in the United States
By Bookmasters